WASTE MANAGEMENT AND VALORIZATION

Alternative Technologies

WASTE MANAGEMENT AND VALORIZATION

Alternative Technologies

Edited by

Elena Cristina Rada, PhD

APPLE ACADEMIC PRESS

Apple Academic Press Inc. 3333 Mistwell Crescent Oakville, ON L6L 0A2 Canada	Apple Academic Press Inc. 9 Spinnaker Way Waretown, NJ 08758 USA

Exclusive worldwide distribution by CRC Press, a member of Taylor & Francis Group

Printed in the United States of America on acid-free paper

International Standard Book Number-13: 978-1-77188-306-1 (Hardcover)

International Standard Book Number-13: 978-1-77188-307-8 (eBook)

Library and Archives Canada Cataloguing in Publication

Waste management and valorization: alternative technologies / edited by Elena Cristina Rada, PhD.

Includes bibliographical references and index.
Issued in print and electronic formats.
ISBN 978-1-77188-306-1 (hardcover).--ISBN 978-1-77188-307-8 (ebook)
1. Refuse and refuse disposal--Technological innovations. 2. Recycling (Waste, etc.)--Technological innovations. I. Rada, Elena Cristina, author, editor

TD791.W38 2016 628.4 C2015-905184-3 C2015-905223-8

Library of Congress Cataloging-in-Publication Data

Waste management and valorization : alternative technologies / Elena Cristina Rada, PhD, editor.

pages cm
Includes bibliographical references and index.
ISBN 978-1-77188-306-1 (alk. paper)
1. Waste minimization. 2. Source reduction (Waste management) 3. Waste products as fuel. 4. Recycling (Waste, etc.) I. Rada, Elena Cristina, editor.

TD793.9.W367 2015 628.4--dc23 2015027255

About the Editor

ELENA CRISTINA RADA, PhD

Elena Cristina Rada, PhD, earned her master's degree in Environmental Engineering from the Politehnica University of Bucharest, Romania; she received a PhD in Environmental Engineering and a second PhD in Power Engineering from the University of Trento, Italy, and the Politehnica University of Bucharest. Her post-doc work was in Sanitary Engineering from the University of Trento, Italy. She has been a professor in the Municipal Solid Waste master's program at Politehnica University of Bucharest, and has served on the organizing committees of "Energy Valorization of Sewage Sludge," an international conference held in Rovereto, Italy, and Venice 2010, an International Waste Working Group international conference. She also teaches seminars in the bachelor, master, and doctorate modules in the University of Trento and Padua and Politehnica University of Bucharest and has managed university funds at national and international level. Dr. Rada is a reviewer of international journals, a speaker at many international conferences, and the author or co-author of about a hundred research papers. Her research interests are bio-mechanical municipal solid waste treatments, biological techniques for biomass characterization, environmental and energy balances regarding municipal solid waste, indoor and outdoor pollution (prevention and remediation) and health, and innovative remediation techniques for contaminated sites and streams.

Contents

Acknowledgment and How to Cite

The editor and publisher thank each of the authors who contributed to this book. The chapters in this book were previously published elsewhere. To cite the work contained in this book and to view the individual permissions, please refer to the citation at the beginning of each chapter. Each chapter was carefully selected by the editor; the result is a book that looks at waste treatment and valorization from a variety of perspectives. The chapters included are broken into five sections, which describe the following topics:

The articles included in the Part I provide an overview of the significant issues within our topic.

- Chapter 1 explains thermal treatments such as vitrification, devitrification, sintering, and room-temperature alkali activation (geopolimerization).
- Chapter 2 summarizes recent developments in waste valorization strategies. The authors focus on food waste in particular.

In Part II, the articles have been selected to represent recent research in treatments and pretreatments.

- Chapter 3 suggests ways to optimize anaerobic digestion as a waste valorization strategy.
- Chapter 4 evaluates the use of ultrasonic pretreatment on olive mill solid waste in the production of methane.

Part III focuses on energy recovery.

- Chapter 5 examines the potential of solid waste as a biomass for energy conversion.
- Chapter 6 studies the thermal degradation behavior of municipal solid waste.

In Part IV, we turn our attention to recycling and reuse as a form of waste valorization.

- Chapter 7 offers research on using coal fly ash from municipal solid waste incinerators for brick manufacturing and other uses.
- Chapter 8 presents the potential for recycled plastic to be used as construction beams.
- Chapter 9 investigates recycling CRT glass by using it to produce durable mortar.
- Chapter 10 examines the challenges that must be overcome by the North American automotive industry in order to recycle automotive plastics.

Finally, in Part V, we consider two additional methodologies for valorizaton:

- Chapter 11 is a case study documenting the ongoing mining of municipal solid waste incinerator ash.
- Chapter 12 provides an analysis of high-temperature slags for the purpose of recovering both waste heat and slag resources.

List of Contributors

F. Andreola
Department of Engineering "Enzo Ferrari", University of Modena and Reggio Emilia, Italy

Rick Arneil D. Arancon
Department of Chemistry, School of Science and Engineering, Ateneo de Manila University, Quezon City, Philippines and School of Energy and Environment, City University of Hong Kong, Hong Kong

L. Barbieri
Department of Engineering "Enzo Ferrari", University of Modena and Reggio Emilia, Italy

R. Borja
Instituto de la Grasa (CSIC), Avenida Padre García Tejero, 4.41012 Sevilla, Spain

L. Bujalance
Instituto de la Grasa (CSIC), Avenida Padre García Tejero, 4.41012 Sevilla, Spain

King Ming Chan
Environmental Science Program, School of Life Sciences, Chinese University of Hong Kong, Hong Kong

Raffaele Cioffi
INSTM Parthenope Research Unit, Department of Engineering, University of Naples Parthenope, Centro Direzionale Is. C4, Naples 80143, Italy

Francesco Colangelo
INSTM Parthenope Research Unit, Department of Engineering, University of Naples Parthenope, Centro Direzionale Is. C4, Naples 80143, Italy

Boukelia Taqiy Eddine
Mechanical Department, Faculty of Engineering, University of Mentouri, Constantine, 25000, Algeria

F. G. Fermoso
Instituto de la Grasa (CSIC), Avenida Padre García Tejero, 4.41012 Sevilla, Spain

Claudio Ferone
INSTM Parthenope Research Unit, Department of Engineering, University of Naples Parthenope, Centro Direzionale Is. C4, Naples 80143, Italy

Mariaenrica Frigione
Department of Innovation Engineering, University of Salento, Via per Arnesano, 73100 Lecce, Italy

Erika Furlani
Dipartimento di Chimica, Università di Udine, Fisica e Ambiente, 108 Via del Cotonificio, 33100 Udine, Italy

Antonio Greco
Department of Innovation Engineering, University of Salento, Via per Arnesano, 73100 Lecce, Italy

Geoffrey John
College of Engineering and Technology, University of Dar es Salaam P.O. Box 35131, Dar es Salaam. Tanzania

Tsz Him Kwan
Environmental Science Program, School of Life Sciences, Chinese University of Hong Kong, Hong Kong

I. Lancellotti
Department of Engineering "Enzo Ferrari", University of Modena and Reggio Emilia, Italy

C. Leonelli
Department of Engineering "Enzo Ferrari", University of Modena and Reggio Emilia, Italy

Carol Sze Ki Lin
School of Energy and Environment, City University of Hong Kong, Hong Kong

Lili Liu
Department of Energy and Resources Engineering, and Management, College of Engineering, Peking University, Beijing 100871, China

Rafael Luque
Departamento de Química Orgánica, Universidad de Córdoba, Campus Universitario de Rabanales, Córdoba, Spain and Department of Chemical and Biomolecular Engineering (CBME), Hong Kong University of Science and Technology, Kowloon, Hong Kong

Alfonso Maffezzoli
Department of Innovation Engineering, University of Salento, Via per Arnesano, 73100 Lecce, Italy

T. Manfredini
Department of Engineering "Enzo Ferrari", University of Modena and Reggio Emilia, Italy

Alessandro Marseglia
CETMA Consortium, S.S.7 Km.706+030, 72100 Brindisi, Italy

A. Martín
Departamento de Ingeniería Química y Química Inorgánica, Facultad de Ciencias, Universidad de Córdoba, Campus Universitario de Rabanales, Edificio C-3, Carretera Madrid-Cádiz, km 396, 14071 Córdoba, Spain

Stefano Maschio
Dipartimento di Chimica, Università di Udine, Fisica e Ambiente, 108 Via del Cotonificio, 33100 Udine, Italy

Francesco Messina
INSTM Parthenope Research Unit, Department of Engineering, University of Naples Parthenope, Centro Direzionale Is. C4, Naples 80143, Italy

Lindsay Miller
Environmental Engineering, University of Windsor, 401 Sunset, Windsor, ON N9B 3P4, Canada

Peter Mtui
College of Engineering and Technology, University of Dar es Salaam P.O. Box 35131, Dar es Salaam. Tanzania

Karoli Njau
School of MEWES, The Nelson Mandela African Institution of Science and Technology, P.O. Box 447, Arusha. Tanzania

Arthur Omari
School of MEWES, The Nelson Mandela African Institution of Science and Technology, P.O. Box 447, Arusha. Tanzania

Alessandra Passaro
CETMA Consortium, S.S.7 Km.706+030, 72100 Brindisi, Italy

C. Ponzoni
Department of Engineering "Enzo Ferrari", University of Modena and Reggio Emilia, Italy

Elena Cristina Rada
Department of Biotechnologies and Life Sciences, Insubria University of Varese, Via G.B. Vico, 46, I-21100 Varese, Italy and Department of Civil Environmental and Mechanical Engineering, University of Trento, Via Mesiano, 77, I-38123 Trento, Italy.

Marco Ragazzi
Department of Civil Environmental and Mechanical Engineering, University of Trento, Via Mesiano, 77, I-38123 Trento, Italy.

B. Rincón
Instituto de la Grasa (CSIC), Avenida Padre García Tejero, 4.41012 Sevilla, Spain

Mahir Said
College of Engineering and Technology, University of Dar es Salaam P.O. Box 35131, Dar es Salaam. Tanzania

Mecibah Med Salah
Mechanical Department, Faculty of Engineering, University of Mentouri, Constantine, 25000, Algeria

Luciano Santoro
Department of Chemical Sciences, University of Naples Federico II, Monte Sant'Angelo Complex, Naples 80126, Italy

Susan Sawyer-Beaulieu
Environmental Engineering, University of Windsor, 401 Sunset, Windsor, ON N9B 3P4, Canada

Katie Soulliere
Environmental Engineering, University of Windsor, 401 Sunset, Windsor, ON N9B 3P4, Canada

Yongqi Sun
Department of Energy and Resources Engineering, and Management, College of Engineering, Peking University, Beijing 100871, China

Edwin Tam
Environmental Engineering, University of Windsor, 401 Sunset, Windsor, ON N9B 3P4, Canada

R. Taurino
Department of Engineering "Enzo Ferrari", University of Modena and Reggio Emilia, Italy

Gabriele Tonello
Dipartimento di Chimica, Università di Udine, Fisica e Ambiente, 108 Via del Cotonificio, 33100 Udine, Italy

Vincenzo Torretta
Department of Biotechnologies and Life Sciences, Insubria University of Varese, Via G.B. Vico, 46, I-21100 Varese, Italy.

Ettore Trulli
Department of Engineering and Environmental Physics, University of Basilicata, via dell'Ateneo Lucano, 10, I-85100 Potenza, Italy.

Simon Tseng
Environmental Engineering, University of Windsor, 401 Sunset, Windsor, ON N9B 3P4, Canada

T. P. Wagner
Department of Environmental Science, University of Southern Maine, Gorham, Maine 04308, USA

Xidong Wang
Department of Energy and Resources Engineering, and Management, College of Engineering, Peking University, Beijing 100871, China and Beijing Key Laboratory for Solid Waste Utilization, Peking University, Beijing 100871, China

Zuotai Zhang
Department of Energy and Resources Engineering, and Management, College of Engineering, Peking University, Beijing 100871, China and Beijing Key Laboratory for Solid Waste Utilization, Peking University, Beijing 100871, China

Introduction

The term "waste valorization" refers to any process that aims to convert waste materials into more useful products by reusing or recycling them (material recovery), or converting them into an energy resource.

On July 2, 2014, the European Commission adopted some proposals aimed at developing a more circular economy in Europe, also promoting recycling in the Member States. The achievement of the new targets for recovery of waste will make Europe more competitive, using materials that would not be wasted but recovered from products at the end of their life. The new targets concern the achievement of 70% municipal solid waste recycling by 2030, of 80% waste packaging recycling by 2030 and of the prohibition of landfilling recyclable waste after 2025. An additional target is the reduction of marine and food waste. Also for the other wastes and discarded biomasses enhanced targets of valorization must be achieved.

Waste valorization usually takes the form of one of the following activities:

- processing waste residue or byproducts into raw materials useful for manufacturing.
- using discarded products as raw materials or energy sources.
- using waste materials in various stages of manufacturing processes.
- adding waste materials to finished products.

Waste valorization is a concept that reflects the growing concerns of the twenty-first century. Until the 1990s, the main objective when dealing with waste management was how to dispose of it or store it. In the 1990s, however, socioeconomic concerns, the depletion of fossil fuels and other raw materials, and the looming crisis of climate change from greenhouse emissions brought about a change in how the world thinks about its waste. Today many market sectors (including transportation, heat and power generation, industrial manufacturing, and construction) are focusing on new technologies to transform low-quality, no-cost materials into high-value products.

Waste valorization is particularly critical in developing countries that are facing energy crises that challenge their economic and social development. Meanwhile, these countries are also struggling to manage their abundant waste materials. Waste valorization offers a solution to both problems. In particular, the valorization strategies in this context should take into account the high percentage of biodegradable materials in municipal solid waste. In this way not only landfilling will be avoided (with favorable effects on the environment), but also the production of compost could support initiatives for bettering the characteristics of soils. Additionally, the valorization of a fraction of biodegradable waste could be coupled with the generation of biogas useful for various domestic and non-domestic purposes. It must be pointed out that in well-organized systems, the valorization concerns not only the collected waste but also the residues generated by the valorization activities.

However, waste valorization should not be regarded as a simple, one-stop answer to the world's energy and environmental problems. At best, it is a complex concept that requires careful study and ongoing research. Because of the need for innovative and complex valorizations strategies, this is an exciting and growing field of study.

The research included in this volume contributes to building a society that regards waste differently from previous generations. Waste is no longer a problem to be buried and burned. Instead, it can be regarded as a resource that can decrease the need for raw materials.

PART I

OVERVIEW

CHAPTER 1

Valorization of Wastes, "End of Waste" and By-Products Using Hot and Cold Techniques

L. BARBIERI, F. ANDREOLA, R. TAURINO, C. PONZONI, T. MANFREDINI, C. LEONELLI, AND I. LANCELLOTTI

1.1 INTRODUCTION

The Framework Directive 2008/98/EC, implemented in Italy by Legislative Decree no. 205/2010, in addition to reporting specific criteria and targets, gave the impression of a new cultural attitude that offers the combination of sustainable use of resources coupled with sustainable management of waste. The adoption of the best available technologies for processing and valorization respects first of all the proximity of the production of the waste. The final aim is the increase of recycling practice to reduce the pressure on the demand for raw materials with a double effect: reuse valuable materials, that would otherwise end up as waste, and reduce energy consumption and emissions of greenhouse gases in the process

Barbieri L, Andreola F, Taurino R, Ponzoni C, Manfredini T, Leonelli C, and Lancellotti I. "Valorization of Wastes, 'End of Waste' and By-Products Using Hot and Cold Techniques" SUM 2014, Second Symposium on Urban Mining Bergamo, Italy; 19 – 21 May 2014.

of extraction and machining. By 2020, we must manage waste as a resource through those instruments and actions that will affect the diffusion of wastes separated collection, promotion of efficient and of high-quality recycling. Contemporary we will assist to the development of markets for the new raw materials obtained by waste recovery and energy recovery limited to the non-recyclable materials regardless of the decrease in per capita and the removal of illegal shipments as well as the landfill.

Although the European order of priority provides for prevention, preparation for re-use, recycling, other recovery (eg energy), disposal, it is not always easy to reduce at source the quantity and hazardous nature of the waste. The latest figures from Eurostat (ISPRA, 2013) on the production of hazardous and non-hazardous waste in the European Union in 2010 estimated a total quantity of approximately 2,515 million tonnes of which about 102 million are dangerous. The valorization of such wastes as well as that of the end of waste and by-products must therefore become a prime target. End of waste criteria specify when a certain waste ceases to be waste and obtains a status of a product (or a secondary raw material). According to Article 6 (1) and (2) of the Waste Framework Directive 2008/98/EC, certain specified wastes shall cease to be waste when they have undergone to a recovery (including recycling) operation and complies with specific criteria to be developed in line with certain legal conditions, in particular:

- the substance or object is commonly used for specific purposes;
- there is an existing market or demand for the substance or object;
- the use is lawful (substance or object fulfills the technical requirements for the specific purposes and meets the exisitng legislation and standards applicable to products);
- the use will not lead to overall adverse environmental or human health impacts.

By-product is a substance or object, resulting from a production process, the primary aim of which is not the production of that item. By-products can come from a wide range of business sectors, and can have very different environmental impacts.

Taking into account the heterogeneity of the above mentioned starting materials and their different nature of hazard, we have hereafter proposed a simple classification of valorization/inertization procedures in "hot" and "cold" techniques. The first group comprises vitrification, devitrification and sintering. Thermal treatments, which might be sped up by the use of electromagnetic irradiation, present the advantage to fix the residue with chemical bonds, changing the microstructure or morphology from hazardous to harmless. In the second group falls stabilization/solidification by both inorganic (cement, lime, clay) and organic (thermoplastic materials, macroencapsulating compounds, asphalts, polymers) reagents.

In this paper will be explained some case studies developed by the writing group of the Department of Engineering "Enzo Ferarri" in Modena (Italy), in collaboration with external industrial partners also. Thermal treatments, such as vitrification, devitrification and sintering, as well as room temperature alkali activation, also known as geopolimerization, will be presented hereafter.

1.2 HOT TECNIQUES

1.2.1 VITRIFICATION

The process of vitrification is gaining increasing importance as a methodology for safer inert hazardous waste. In this field it is worth mentioning the invaluable experience gained with the inerting of high-level radioactive waste, which since the 70s, especially in France, the U.S. and Britain, has led to the identification of suitable compositions of glasses as container ideal to ensure, in times of the order of the 100,000 years, no leaching of radioactive isotopes with a high half-life in them "solubilized" at the atomic level. Until a few years ago, various factors combined to make unattractive the vitrification process: the non cost-effectiveness of the process (the costs of which were related to the processing of high-temperature fusion, the need to introduce corrective, sometimes to a considerable extent, to obtain glasses stable or to lower the temperature of processing, etc.), the lack of adequate legislation, the easy availability of landfills (more or less legal), the widespread tendency to underestimate the risks to health and the environment related to outright abandonment or burial of waste, etc. Today the situation is changing. Laws impose severe restrictions on the making inert preventive hazardous waste; landfill permitted more and more scarce, even for the hostilitity of people to start new ones, and consequently the cost of landfilling rises more and more. It is also forming an ecological consciousness, which leads to fighting against illegal landfills and illegal organizations and illegal organizations waste disposal. The vitrification process, thanks to the evolution of construction of the smelting furnaces and the development of new systems cheaper, is starting to become competitive to other disposal systems, especially considering that it is the only one able to provide absolute guarantees inerting long term and to allow large, sometimes very large, volume decreases. For vitrification, whether the waste contains little or no amount of substance of "forming glassy network", it takes appropriate additions and corrections to the composition of the waste, to be made with scrap glass. The final product of a successful vitrification can be used for road surfacing, for dams or embankments for consolidation works or foundations, or added in the manufacture of cement or ceramic.

Vitrification process is based on raw materials (carbonates, oxides, etc.) heating at high temperatures (1200-1700°C). This treatment creates a melted product, which, submitted to a rapid cooling, produces glass. As regards wastes management, vitrification possess several advantages such as:

- process flexibility, thanks to the possibility to incorporate into glass structure any element of periodic table (included wastes which usually have a very complex chemical composition);
- destruction of organic pollutants, such as dioxins;
- obtainment of a material with high chemical stability;
- waste volume reduction;
- possibility to use the final harmless vitrified material.

Some vitrification processes occur naturally, i.e. glasses produced by volcanic eruptions, such as obsidians (Na_2O-K_2O-Al_2O_3-SiO_2) and basaltic rocks (CaO-MgO-Al_2O_3-SiO_2), and are basically inert materials even when they contain toxic elements because these are embedded in an extremely stable glassy matrix. Being vitrification an energy-intensive process with costs related, its use is better justified if a high-quality vitrified product with optimised properties, for example glass-ceramic, can be fabricated.

1.2.2 DEVITRIFICATION

The most effective way to improve the properties of the vitrified material is the transformation into glass-ceramic. This is a fine-grained polycrystalline material (typically from 50% to 98% of the volume may be crystalline) formed when a glass of suitable composition is heat treated and hence undergone to controlled crystallisation (phenomenon named devitrification).

The term devitrification refers to a combination of two processes: nucleation and crystal growth. In particular, nucleation is homogeneous if nuclei form themselves spontaneously within the melt. Heterogeneous nucleation refers to nuclei which are generated on a pre-existing surface, such as that due to an impurity, crucible wall, etc. If no nuclei are present, crystal growth cannot occur and the material will remain amorphous. If some nuclei are present, but no crystal growth has occurred, the extremely small size and low volume fraction of the nuclei prevents their detection, so that the solid is still considered a glass.

Glass-ceramics can be obtained by petrological or glass-ceramic process. In the first case, the melt is cooled slowly from melting to room temperature into the original mould. During the cooling stage, both nucleation and crystal growth are accomplished. In the second case, the starting glass is obtained via melting and rapid quenching. After obtaining the glass, the material is heat-treated in one or, more frequently, two stages during which nucleation take place and crystal growth starts.

1.2.3 SINTERING

Sintering is the bonding together of particles when heated to high temperatures.

Liquid phase sintering (LPS) is a subclass of the sintering processes involving particulate solid along with a coexisting liquid during some part of the thermal cycle. Presence of liquid phase during sintering gives this process an advantage over solid state sintering processes providing both a capillary force and a transport medium that leads to rapid consolidation and sintering. The capillary attraction due to the wetting liquid gives rapid compact densification without the need of an external force. As the starting material is in a powder form, a near net shape fabrication is possible, and complete melting is not necessary. All these features make LPS a very attractive fabrication process for commercial production. LPS is a widely used fabrication process for high performance metallic and ceramic (glass–ceramic) materials. It is particularly useful for high melting point materials for which fabrication by melting is not feasible.

Limiting our attention to ceramics, defined as every object made of inorganic raw materials, not metallic, shaped at room temperature and consolidated by heating, advanced ceramics are products characterized by high structural and functional performances (carbides (SiC), pure oxides (Al_2O_3), nitrides (Si_3N_4), non-silicate glasses and many others).

Traditional ceramic industry includes the manufacture of pottery, porcelain, building ceramics (e.g., bricks, tiles, stoneware). Ceramics are prepared from malleable, earthy materials, such as clay, that become rigid at high temperatures. Besides clays, which confer cohesion and plasticity, and feldspar as melting agent, ceramic pastes also include inert materials that provide structural support, necessary to retain shape during drying and firing. Quartz (SiO_2) is the most commonly used inert material and is usually supplied either as sand or schist. It is also used in the glazing of ceramic bodies in combination with plasticer and melting agents. When a glaze layer is applied to the ceramic body and then fired, the glaze ingredients melt and become glass-like. During thermal treatment (sintering at a maximum temperature around 1200°C for less than 1 hour) the aforementioned raw materials are transformed into crystalline silicate phases and glassy matrices capable of fixing the wastes. The ceramic industry represents an important reference point for the recycling or re-use of several waste types; for example, integrated-gasification gas-combined cycles slag (Acosta et al., 2002), and windshield glasses (Mortel and Fuchs, 1997) for brick manufacture; municipal solid wastes incinerator fly ash and granite sawing residues (Hernàndez-Crespo and Rincòn, 2001) for porcelainized stoneware, and the used catalyst from the fluidized-catalyst cracking units involved in petroleum refining in the preparation of ceramic frits (Escardino et al., 1995).

In Italy, manufacturing of wall and floor ceramic tiles is an important part of the economy, primarily in an area of 300 km^2 in the north of the country, that is, in the provinces of Modena and Reggio Emilia, where 80% of tiles manufactured are produced nationally. The raw materials' (about 56% imported from other countries such as Ukraine, Turkey, Germany and France) costs are important and, therefore, the complementary use of alternative raw materials is of significant interest in the ceramic sector.

Besides tiles, brick manufacture can be an interesting potential option for the recovery of value added materials. In this process, raw materials preparation is made by dry grinding, the moulding is performed by extrusion and the firing cycle is characterized by low temperatures ($T <$ 1000°C) leading to a porous product by solid state sintering, due to the absence of a liquid phase responsible of the densification; therefore an increase of porosity, sometimes caused by degasable components present in wastes, does not represent an undesired factor. The bricks manufacture is controlled by the ratio of relatively refractory minerals that maintain the shape of a ceramic body to the easily melted materials that fuse and produce a steel-hard brick. A further consideration is having enough plasticity for good extrusion and enough clay minerals for good green (unfired) strength. Bricks present a production cycle particularly simple, tolerance regard to technical parameters greater than other productive typologies, and the property of the fired product to absorb heterogeneous materials. For these reasons bricks represent a concrete possibility to reuse by-products of different nature, also in high amount.

1.2.4 ALKALI ACTIVATION/GEOPOLIMERIZATION

Geopolymers are a type of inorganic polymers that can be formed at room temperature by using industrial waste or by-products as source materials to form a solid binder that looks like and performs a similar function to cement. Differently from cement, geopolymers are made from aluminium and silicon, instead of calcium and silicon. The alkali activation of aluminosilicate powders (either from industrial waste, by-products, "end-of-waste" or eventually natural rocks cutting sludges and debris) is a complex chemical reaction simplified in three steps:

1. Alkaline depolymerization of the poly(siloxo) layer of the starting aluminosilicate powder;
2. Formation of the ortho-sialate $(OH)_3$-Si-O-Al-$(OH)_3$ molecule;
3. Polycondensation into higher oligomers and polymeric 3D-networks.

In a word, geopolymerization forms aluminosilicate frameworks that are similar to those of rock-forming minerals. The most readily available raw materials containing aluminium and silicon are fly ash and metallurgical slag, but also other type of wastes could be suitable, such as incinerator bottom ash, ladle slag, etc.

When used as binder, geopolymer can be used in applications to fully or partially replace Portland cement with environmental and technical benefits, including an 80–90% reduction in CO_2 emissions and improved resistance to fire and aggressive chemicals. Unlike other industrial processes, indeed, geopolymerization does not require high temperature thermal treatments, nor the use of carbonate-based raw materials or expensive chemical reagents.

1.3 CASE STUDIES

1.3.1 *VITRIFICATION/DEVITRIFICATION*

Glass cullet deriving from different sources (separated urban collection (SUC), waste of electrical end electronic equipments (WEEE)) were used as vitrifing agent for ashes of thermal power plant, municipal solid waste incinerator (MSWI) and steelworks, lagoon sludge and rice husk ash.

Ternary mixtures containing power plant ashes (up to 50 wt%), SUC glass cullet and dolomite as by-product were melted and transformed into glass-ceramics (by appropriate thermal treatments) with dendritic and/or acicular microstructures of diopside/augite and wollastonite and improved mechanical properties (Hv = 5-6.9 GPa; E = 59-101 GPa e K_{IC} = 1.7-1.8

MPam$^{1/2}$) with respect to the parent glasses (Barbieri et al., 1997). On the other hand, binary compositions without dolomite originated very stable glasses showing no significant attack either in water or acid (Barbieri et al., 1999). Time-temperature-transformation curves (obtained by XRD data) showed that crystallisation kinetics and critical cooling rates are in the 12-42°C/min interval (Barbieri et al., 2001).

Municipal incinerator bottom ash and sludge excaved from the Venice lagoon were vitrified at 1350-1500°C by themselves or after the addition of SUC glass cullet. The obtained glasses possessed a good chemical durability, and were drawn into fibres at various temperatures without crystallizing. The fibres possess a good tensile strength (a maximum value of 1.6 GPa was obtained), although lower than that of conventional fibres of glass, because their surface was not protected by a polymeric layer after drawimg. The elastic modulus value, however, was comparable to that of E glass fibres, and increased with the solid waste (ash or sludge) content in the glass, while it did not depend on the drawing temperatures (Scarinci et al., 2000).

By mixing different amount of MSWI bottom and fly ashes with others inert materials, such as SUC glass cullet and feldspar waste, vitrifiable mixtures with tendency to crystallise if subjected to thermal treatment were tailored. The glasses obtained showed comparable properties to those of commercial soda-lime glasses and low leachability of contaminants (Andreola et al., 2008). High contents of bottom ash mixed with glass cullet generate the formation of bulk and sintered glass-ceramics favouring the growth of the crystalline volume fraction and the formation of pyroxene, anorthite and wollastonite (thermal analysis, X-ray powder diffraction and hot-stage microscopy were employed). Because the nucleation mechanism starts from the surface and sintering occurs before the crystallisation, all the compositions umidified with a water solution are easily sinterable in dense materials at temperatures relatively low, around 850°C, so rendering the process economically advantages (Barbieri et al., 2000). The sinter-crystallization ability was also studied as function of ash particle size (coarse and fine). The phase formation was estimated by DTA and XRD, while the sintering process was evaluated by optical dilatometry, linear shrinkage and water absorption. The porosity variations were estimated by density measurements. The microstructure and morphology of the glass-ceramics were observed by scanning electron microscopy. This integrated experimental approach together with theoretical study (by the methods of Ginsberg, Raschin–Tschetveritkov and Lebedeva) permitted to establish a better sinter-crystallization ability for the glass obtained from coarse ash fraction (Maccarini Schabbach et al., 2011). Crystallization kinetics were thoroughly investigated by thermal studies performed both in air and argon atmospheres. The investigated composition is characterized with a high crystallization trend and formation of pyroxene solid solutions and melilite solid solutions. Due to additional nucleation process and lower viscosity (because of the lack of Fe^{2+} oxidation) the phase formation in inert atmosphere is accelerated and is carried out at lower temperature. In the interval 800–900 °C the densification in both atmospheres is inhibited by the intensive phase formation. However, after increasing the sintering temperature up to 1120–1130°C secondary densification is carried out, resulting in material with zero water absorption, low closed porosity and high crystallinity. Some decreasing of sintering temperature and finer crystal structure are predicted at densification in inert atmosphere (Karamanov et al., 2014).

Chemically inert and differently coloured glasses and glass-ceramics were prepared by inserting up to 10 wt% steel plant fly ash into different kinds of inorganic matrices (municipal incinerator grate ash, glass cullet and a low-cost Ca-Mg-aluminosilicate devitrifiable glass) without changing substantially the thermal and mineralogical behaviour (Barbieri et al., 2002).

The vitrification treatment has been successfully exploited as a solution for the disposal of polluted dredging spoils from the industrial area close to the Venice lagoon. The addition of 20 wt% of SUC glass cullet to the calcined sediments in the vitrification batch provides a suitable chemical composition for the production of an inert glass, despite the compositional variations of the sediments. The obtained waste glass, after being finely ground, has been employed (i) as a raw material for the manufacture of sintered glass-ceramics, by cold pressing and single-step sintering at about 940°C, and (ii) as sintering additive (the maximum addition being 10 wt%) for the manufacture of traditional red single firing ceramic tiles, with a maximum firing temperature of 1186°C. Both applications have proved to be promising: in the first case, the sintered glass ceramic product exhibits notable mechanical properties (bending strength >130 MPa, H_V≈6.5 GPa); in the second case, the addition of waste glass does not modify substantially the investigated physical and mechanical properties of the traditional product (water absorption, linear shrinkage, bending strength, planarity) (Brusatin et al., 2005).

A commercial ceramic glaze composed by both olivine (magnesium iron silicate, $(Mg,Fe)_2SiO_4$) and commercial frits, rich in lead (about 30 wt%), was reformulated by using secondary raw materials (CRT cone glass and MSWI post-treatment bottom ash before and after vitrification). The waste-based products were characterized and, compared to the standard glaze, showed better acid resistance, comparable aesthetic characteristics and slightly lower stainless resistance. Environmental benefits were obtained by saving natural raw material (olivine), by reducing lead percentage in the proposed formulations (from around 30 to 5 wt%), by energy saving (for the avoided use of commercial frits) and by reducing lead content in the new compositions (Schabbach et al., 2011). Another ceramic glaze was successfully obtained by susbstituting more than 80 wt% of the glassy component (frit) of a commercial double firing ceramic glaze. A decrease in the environmental impact was also demonstrated by LCA method with respect to the commercial glazed obtained starting by virgin raw materials (Andreola et al., 2007).

Cathode ray tubes glass added with alumina and dolomite were melted around 1500°C and transformed into glass-ceramics applying different thermal cycles (900-1100°C and 0.5-8 hours soaking time). By elaborating thermal, mineralogical and microstructural results was carried out that a good crystallization degree is achieved around 1000°C and with a glass percentage around 50-75 wt% if an amount of 40-45 wt% of CaO and MgO (in the form of dolomite) is introduced. Mainly alkaline and alkaline-earth silicates and aluminosilicates develop with a possible surface mechanism (Andreola et al., 2005).

Fluorescent lamps or cathode ray tubes glass and packaging scrap glass has been used for the matrix of glass foams. Other wastes or by-products, egg shells and cutting sludge from treatment of glass polishing (containing $CaCO_3$ in percentage of 95-97 and 7, respectively) were used as foaming agents. These last, introduced in the matrix in media from 1 to 5 wt%, permitted the obtainement of glass foams with density values from 0.24

to 0.69 g/cm^3, porosity (81 ÷ 87%) and presenting interesting mechanical properties such as compressive strangth up to around 15 MPa for samples obtained at temperatures from 650 to 800°C for 45, 30 or 15 minutes (Andreola et al., 2012; Fernandes et al., 2013).

Rice husk ash was successfully used as raw material (silica source) to develop forsterite–nepheline glass-ceramic. An original glass was formulated in the base system MgO-Al_2O_3-SiO_2 with addition of B_2O_3 and Na_2O to facilitate the melting and poring processes. By a sintercrystallization process, glass-ceramics with nepheline ($Na_2O \cdot Al_2O_3 \cdot SiO_2$) as major crystalline phase was obtained in the temperature interval 700-950°C and forsterite ($2MgO \cdot SiO_2$) at temperatures above 950°C (Martín et al., 2011). Deeping on crystallisaztion activation energies and morphology of the materials obtained were also performed (Martín et al., 2013). Regarding technological features, the sintered materials showed bending strength values and Mohs hardness higher with respect to commercial glass-ceramics like Neoparies®. Other properties as water absorption (0.5%) allowed to classify these materials into the Group BIa characteristic of high sintered ceramic tiles according to European Standard rule (Andreola et al., 2013).

1.3.2 SINTERING

Selected urban collection (SUC) glass cullet has been used to prepare a unique new materials where its content could be as high as 90 wt% (UNIMORE-Ingrami, 2011). The glass waste after being ground to powder with different grain size, has been shaped with different forming technique in slabs, tiles, vases, and other artistic forms before being subjected to thermal cycle. The final firing temperature of the new ceramic formulation has been lowered by 250-300°C below common porcelain stoneware body. Peculiar surface features and colors have been investigated and a number of niche applications specially in the design objects have been found.

Additionally SUC or WEEE glass cullet, ceramic sludges, MSWI and rice husk ash (RHA) were tested in the formulation of tiles or bricks.

MSWI bottom and fly ashes were inserted into ceramic bodies. Good results were obtained with bottom ash, in particular if pre-vitrified, successfully inserted as sintering promoters into porcelain stoneware ceramic body in percentage ranging from 5 to 10 wt%, respectively without significantly changing technical parameters of the final products and with good leaching answer. Nevertheless, a little increase of the porosity in proportion to the quantity of waste introduced, produces a lowering of density, shrinkage, resistance to stain and white point (Andreola et al., 2001; Andreola et al., 2002; Barbieri et al., Waste Manage. 2002; Rambaldi et al., 2010). Due to the presence of cromophore oxides inside the bottom ash, better results from a chemical and technological point of view were reached using a red ceramic body for glazed wall tiles. Higher densificaton, constance or light increase of mechanical and mineralogical properties were detected with colored body with respect to the porcelain stoneware one (Andreola et al., Ceram. Inf. 2001).

The partial replacement of the melting component allows the addition up to a 5-10 wt% glass of CRT into porcelain stoneware ceramic body without modifying the parameters of the sintering. The microstructure and the mechanical properties of the materials obtained compared with the standard composition are very similar. Furthermore, the CRT was successfully used up to 15 wt% to completely replace feldspar and inert components in a traditional ceramic mix. The samples obtained have mechanical and technological properties similar to those of commercial floor and wall products (Andreola et al., Ceram. Int. 2008).

Glazing and polishing ceramic sludges were used as raw materials in commercial formulations of bricks and tiles using traditional thermal cycles. 0-20 wt% polishing sludge-added bricks showed aesthetic (efflorescences, colour) and technological (water absorption, weight loss, shrinckage, comprehession resistance) properties comparable to those of the standards, within the limits of industrial tolerance. The partial substitution of Na-feldspars with polishing sludge in porcelain stoneware ceramic body produces a decrease of the sintering temperature maintaining the products properties constants. Moreover, the addition of the same waste into a single firing ceramic body, by decreasing the porosity improves the mechanical resistance of the tiles (Andreola et al., 2006).

Rice husk ash (RHA) ranging from 5 to 20 wt% was used as silica resource mixed with an industrial clay mixture and 20% of water to produce bricks. Laboratory brick samples were obtained by extrusion, and were fired in an industrial kiln with a thermal cycle (24 hrs total time,Tmax:

960°C and 6 hrs soaking time). The dried and fired specimens were characterized following the technical rules and compared to the standard one (clayed materials). This research has revealed that RHA behaves as a raw materials with high silicate content which have mainly plasticity reducing effect on the brick bodies. RHA is compatible in the formulation of bricks up to 5 wt% while not worsening the mechanical and structural properties compared to the standard one. For that percentage, the addition of RHA results satisfactory from mechanical point of view and the data obtained are in accordance to the recommended values for floor (10 MPa). Bricks containing higher amounts of RHA could be used in building manufacturing (light weighted faced load bearing walls) where moderate strengths and penetration protection (porosity/permeability) are required (Andreola et al., 2007; Bagnoli, DegreeThesis 2008; Bondioli et al., 2010).

1.3.3 ALKALI ACTIVATION/GEOPOLYMERIZATION

Different kinds of not dangerous wastes can be used as alluminosilicate source for producing geopolymers, while hazardous wastes can be inertized in the geopolymer matrix. In particular, the wastes considered by the authors as raw materials for geopolymers are incinerator bottom ash and ladle slag.

MSWI bottom ash has been used in the present study for producing dense geopolymers containing high percentage (50-70 wt%) of ash. The amount of potentially reactive phase in the ash has been determined by means of test in NaOH. The final properties of geopolymers prepared with or without taking into account this reactive fraction in the formulation have been compared. The results showed that due to the presence of both amorphous and crystalline fractions with a different degree of reactivity, the incinerator bottom ash exhibits significant differences in the Si/Al ratio and microstructure (more homogeneous) when reactive fraction is considered. In view of sustainability where cities can be considered as urban mine, the results reported in this paper show as a waste or an "end of waste" material can be a valuable resource for obtaining new materials by saving conventional raw materials and energy (Lancellotti et al., 2013).

With the aim to further increase the type of waste to be used as precursor and to promote a new recycling route, alkali activated materials based on partial substitution of metakaolin with ladle slag, deriving from the refining process of steel produced by arc electric furnace technology, are considered. In particular, being ladle slag rich in Ca-containing crystalline phases, its effect on the consolidation process has been investigated. The results of this research can be summarized as follows:

- ladle slag, when alkali activated, participates in the consolidation process even if the crystalline phases characteristic of the slag do not react completely, thus remaining as un-soluble fraction dispersed in amorphous matrix;
- ladle slag content strongly influences the total open porosity and pore size distribution. Indeed, an optimization in terms of critical pore radius has been detected;
- joint activation of metakaolin and ladle slag allows the formation of different gels which differ for chemical nature and structures. Besides C–S–H and 3D aluminosilicate network, C–A–S–H gels with different content of calcium and sodium have been detected. An ion exchange mechanism between calcium of ladle slag and sodium of 3D aluminosilicate network has been envisaged. From a morphological point of view the homogeneity of the gels becomes lower when very high amount of ladle slag is present (Bignozzi et al., 2013).

For the inertization of dangerous wastes, authors worked on MSWI fly ashes coming from electrofilter and fabric filters for gases depuration. Municipal solid waste incinerators, indeed, every year produce tons of fly ashes which, differently from coal fly ashes, contain large amounts of toxic substances such as heavy metals. Geopolymerization is proposed with the purpose to bond physically and chemically incinerator fly ashes in a solid matrix, in order to reduce pollutant mobility. The chemical stability of geopolymers with Si/Al ratio of 1.8-1.9 and Na/Al ratio of 1.0, synthesized by alkali activation of metakaolin and the addition of 20 wt% of two different kinds of IFA, is presented. The concentration of the alkaline solution, water to solid ratio and curing process have been optimized. The room temperature consolidation of MSWI fly ashes-containing geopolymers has been tested for leachability in water for one day, accordingly to EN 12457 regulation and extended to 7 days to increase the water attack on solid granules. Leachable metals in the test solution, determined by ICP_AES, fall within limit values set by regulation for non-dangerous waste landfill disposal. Geopolymeric matrix evolution with leaching time has been also evaluated in terms of pH and electrical conductivity increase in solution. The preliminary results obtained highlight that the waste is successful incorporated into the inorganic polymeric matrix and the absence of a washing pre-treatment reduces the use of water. The results are more encouraging when lower solubility in the starting MSWI fly ashes is encountered. For high soluble fly ash containing geopolymers, the leaching of heavy metals is stronger and the overall geopolymeric matrix is weaker. Nevertheless an impressive incorporation of soluble salts, i.e. chlorides, carbonate, and in minor extent sulphates, has been noticed after a prolonged 7 days leaching test (Lancellotti et al., 2010).

1.4 CONCLUSIONS

This work proposes sustainable solutions in residential and non-residential buildings by using ceramic materials and geopolymers obtained by substitution of virgin raw materials with different kinds of waste into the body. The residues used in this paper derived from various industrial sectors (ceramic, agricultural, metallurgical, electronic) and from separate collection and incinerator plant indicating how a proper waste management collection can favour building materials industry.

Hot and cold treatments have proved valuable techniques to recovery different kinds of wastes to obtain new materials or materials with similar properties of the commercial ones (Table 1). Additionally we also propose these techniques for inertization of hazardous wastes.

The recycling of these wastes may present clear advantages from economical and technological point of view; the reduction of costs relates to use of alternative raw materials; the reduction of consume of virgin raw materials and the saving energy in cold treatments.

It has been also pointed out how a correct scientific approach taking into account the chemical, physical, and thermal properties of the waste jointly with the peculiarities of the hot/cold technique chosen is necessary to achieve the desired final product.

TABLE 1: Different kinds of residues hot and cold treated and the products obtained.

Waste	Treatment	Product
MSWI ash/dolomite/SUC glass cullet	Vitrification	Glass-ceramic
MSWI bottom ash/Venice lagoon sludge/SUC glass cullet	Vitrification	Glass fibre
MSWI bottom and fly ashes/ SUC glass cullet/feldspar waste	Vitrification	Glass
MSWI bottom ash/SUC glass cullet	Vitrification	Glass-ceramic
Steel plant fly ash/MSWI bottom ash/glass cullet/Ca-Mg-aluminosilicate devitrifiable glass	Vitrification	Glass and glass-ceramic
MSWI bottom ash/CRT cone glass	Vitrification	Ceramic glaze (frit)
Cathode ray tubes glass/ alumina/dolomite	Vitrification/Devitrification	Glass-ceramic
Fluorescent lamps or CRT glass/packaging scrap glass/ waste foaming agent	Vitrification	Foam glass
Rice husk ash (RHA)	Vitrification/ Sinter-crystallization	Glass-ceramic
MSWI bottom and fly ashes (5 -10 wt%)	Sintering	Glazed wall tile
CRT glass (5-10 wt%)	Sintering	Porcelain stoneware tile
Glazing and Polishing sludges (up to 20 wt%)	Sintering	Brick
	Sintering	Brick
Polishing sludges (up to 15 wt%)	Sintering	Ceramic tile
RHA (5 wt%)/clay mixture	Sintering	Bricks
MSWI bottom ash (50-70 wt%)	Alkali activation	High dense geopolymer
Ladle slag from refining steel produced by arc electric furnace	Alkali activation	Geopolymer
MSWI fly ashes (20 wt%)	Alkali activation/inertization	Inertized material

REFERENCES

1. Acosta A., Iglesias I., Aineto M., Romero M., Rincòn J. Ma. (2002). Utilisation of IGCC slag and clay steriles in soft mud bricks (by pressing) for use in building bricks manufacturing. Waste Manag., vol. 22, 887-891.
2. Andreola F., Barbieri L., Lancellotti I. (2001). Problematiche e prospettive di valorizzazione di scorie di inceneritore urbano nel settore ceramico. Ceram. Inf. vol. 408, 649-652.
3. Andreola F., Barbieri L., Corradi A., Lancellotti I, Manfredini T. (2001). The possibility to recycle solid residues of the municipal waste incineration into a ceramic tile body. J. Mater. Sci. vol. 36, 4869-4873.
4. Andreola F., Barbieri L., Corradi A., Lancellotti I, Manfredini T. (2002). Utilisation of municipal incinerator grate slag for porcelanized stoneware tiles manufacturing. J. Eur. Ceram Soc. vol. 22 (9,10), 1457-1462.
5. Andreola F., Barbieri L., Corradi A., Lancellotti I, Falcone R., Hreglich S. (2005). Glass-ceramics obtained by the recycling of end of life cathode ray tubes glasses. Waste Manage. vol. 25, 183-189.
6. Andreola F., Barbieri L., Lancellotti I, et al. (2006) Treatment and valorisation of polishing and glazing ceramic sludge in Innovative technologies and Environmental impacts in waste management, pp 152-155, Ed. L. Morselli et al., Maggioli Editore. ISBN:88-387-3645-6.
7. Andreola F., Barbieri L., Corradi A., Ferrari A. M,. Lancellotti I, Neri P. (2007). Recycling of EOL CRT glass into ceramic glaze formulations and its environmental impact by LCA approach. Int. J. LCA vol.12, n.6, 148-454.
8. Andreola F., Barbieri L., Lancellotti I, Hreglich S.., Morselli L., Passarini F., Vassura I. (2008). Reuse of incinerator bottom and fly ashes to obtain glassy materials. J. Hazard. Mater., vol.153, 1270-1274.
9. Andreola F., Barbieri L., Bondioli F.,. Ferrari A. M, Manfredini T. (2007). Valorization of Rice Husk Ash as Secondary Raw Material in the Ceramic Industry. Proc. 10th ECerS Conf., Göller Verlag, Baden-Baden, 1794-1798, ISBN: 3-87264-022-4.
10. Andreola F., Barbieri L, Karamanova E., Lancellotti I., Pelino M. (2008). Recycling of CRT panel glass as fluxing agent in the porcelain stoneware tile production. Ceram. Int. vol. 34, 1289-1295.
11. Andreola F., Barbieri L, Giuranna D., Lancellotti I. New eco-compatible materials obtained from WEEE glass in Atti The ISWA world solid waste congress 2012, chiavetta USB (poster 153) (Firenze, 17-19 Settembre 2012.
12. Andreola F., Martín M. I., Ferrari A.M., Lancellotti I., Bondioli F., Rincón J. Ma., Romero M., Barbieri L. (2013). Technological properties of glass-ceramic tiles obtained using rice husk ash as silica precursor. Ceram. Intern. vol.39, 5427-5435.
13. Barbieri L., Manfredini T, Queralt I., Rincòn J. Ma., Romero M. (1997). Vitrification of fly ash from thermal power stations. Glass Technol., vol. 38 n.5, 165-170.
14. Barbieri L., Lancellotti I., Manfredini T., Queralt I., Rincòn J. Ma., Romero M. (1999). Design, obtainment and properties of glasses and glass-ceramics from coal fly ash. Fuel, vol. 78, 271-276.

15. Barbieri L., Corradi A., Lancellotti I. (2000). Bulk and sintered glass-ceramics by recycling municipal incinerator bottom ash. J. Europ. Ceram. Soc. vol. 20 n.10, 1637-1643.
16. Barbieri L., Lancellotti I, Manfredini T., Pellacani G. C., Rincòn J. Ma., Romero M. (2001). Nucleation and crystallization of new glasses from fly ash originating from thermal power plants. J. Am. Ceram. Soc. vol. 84 n.8, 1851-1858.
17. Barbieri L., Corradi A., Lancellotti I, Manfredini T. (2002) Use of municipal incinerator bottom ash as sintering promoter in industrial ceramics. Waste Manage. vol. 22 n.8, 859-863
18. Barbieri L., Corradi A., Lancellotti I. (2002). Thermal and chemical behaviour of different glasses containing steel fly ash and their transformation into glass-ceramics. J. Eur. Ceram. Soc. vol. 22 n.11, 1759-1765.
19. Bagnoli N. Degree Thesis. 2008. University of Modena and Reggio Emilia.Tutors: Bondioli F., Barbieri L., Andreola F.
20. Bignozzi M. C., Manzi S., Lancellotti I., Kamseu E., Barbieri L., Leonelli C. (2013). Mix-design and characterization of new geopolymer system based on metakaolin and ladle slag. Apply Clay Science. vol. 73, 78–85.
21. Bondioli F., Barbieri L., Andreola N. M., Bonvicini M. (2010). Agri-food waste: an opportunity for the heavy clay sector. Brick World Review vol 1, 34-40.
22. Brusatin G., Bernardo E., Andreola F., Barbieri L., Lancellotti I., Hreglich S. (2005). Reutilization of waste inert glass from the disposal of polluted dredging spoils by the obtainment of ceramic products for tiles applications. J. Mater. Sci. vol. 40, 5259-5264.
23. Escardino A., Amoros J. L., Moreno A. and Sanchez E. (1995). Waste Manage. vol. 13 n.6, 569-577.
24. Fernandes H. R., Andreola F., Barbieri L., Lancellotti I., Pascual M. J., Ferreira J. M. F. (2013). The use of egg shells to produce Cathode Ray Tube (CRT) glass foams. Ceram. Intern. vol. 39 n.8, 9071-9078.
25. Hernàndez-Crespo M. S., Rincòn J. Ma. (2001). New porcelainized stoneware materials obtained by recycling of MSW incinerator fly ashes and granite sawing residues.Ceram. Int. vol.27, 713-720.
26. ISPRA – National Report of Municipal Waste 2013. Roma, Italy.
27. Karamanov A., Maccarini Schabbach L., Karamanova E., Andreola F., Barbieri L., Ranguelov B., Avdeev G., Lancellotti I. (2014). Sinter-crystallization in air and inert atmospheres of a glass from pre-treated municipal solid waste bottom ashes. J. Non-Cryst. Solids vol.389, 50-59.
28. Lancellotti I., Kamseu E., Michelazzi M., Barbieri L., Corradi A., Leonelli C. (2010). Chemical stability of geopolymers containing municipal solid waste incinerator fly ash. Waste Manage. vol.30, 673-679.
29. Lancellotti I., Ponzoni C., Barbieri L., Leonelli C. (2013). Alkali activation processes for incinerator residues management. Waste Manage.vol. 33 n. 8, 1740-17499.
30. Maccarini Schabbach L., Andreola F., Karamanova E., Lancellotti I., Karamanov A., Barbieri L. (2011). Integrated approach to establish the sinter-crystallization ability of glasses from secondary raw material. J. Non-Cryst. Solids vol.357, 10-17.

31. Martín M. I., Rincón J. Ma., Andreola F., Barbieri L., Bondioli F., Lancellotti I., Romero M. (2011). Materiales vitrocerámicos del sistema MgO-Al2O3-SiO2 a partir de ceniza de cáscara de arroz. Bol. Soc. Esp. Cer. Vid. vol. 50 n.4, 169-176.
32. Martín M. I., Rincón J. Ma., Andreola F., Barbieri L., Bondioli F., Lancellotti I., Romero M. (2013). Crystallisation and microstructure of nepheline-forsterite glass-ceramic. Ceram. Intern. vol. 39, 2955-2966.
33. Mortel H., Fuchs F. (1997). Recycling of windshield glasses in fired brick industry. Key Engineering Materials Trans Tech Publications, Switzerland. 132-136: 2268-2271.
34. Rambaldi E., Esposito L., Andreola F., Barbieri L., Lancellotti I., Vassura I. (2010). The recycling of MSWI bottom ash in silicate based ceramic. Ceram. Int. vol. 36, 2469-2476.
35. Scarinci G., Brusatin G., Barbieri L., Corradi A., Lancellotti I., Colombo P., Hreglich S., Dall'Igna R. (2000). Vitrification of industrial and natural wastes with production of glass fibers. J. Eur. Ceram. Soc. vol. 20 n.14-15, 2485-2490.
36. Schabbach L. M., Andreola F., Lancellotti I., Barbieri L. (2011). Minimization of Pb content in a ceramic glaze by reformulation the composition with secondary raw materials. Ceram. Int. vol. 37, 1367-1375.
37. UNIMORE (Barbieri L., Leonelli C., Andreola F., Reggiani E.) – Ingrami M. (2011). Materiale a base vetrosa per la produzione di manufatti ceramici e metodo per la sua preparazione. Patent n. 0001404410.

CHAPTER 2

Advances on Waste Valorization: New Horizons for a More Sustainable Society

RICK ARNEIL D. ARANCON, CAROL SZE KI LIN,
KING MING CHAN, TSZ HIM KWAN, AND RAFAEL LUQUE

2.1 INTRODUCTION

Climate change, energy crisis, resource scarcity, and pollution are major issues humankind will be facing in future years. Sustainable development has become a priority for the world's policy makers since humanity's impact on the environment has been greatly accelerated in the past century with rapidly increasing population and the concomitant sharp decrease of ultimate natural resources. Finding alternatives and more sustainable ways to live, in general, is our duty to pass on to future generations, and one of these important messages relates to waste. Waste from different types (e.g., agricultural, food, industrial) is generated day by day in extensive quantities, generating a significant problem in its management and disposal. A widespread feeling of "environment in danger" has been present everywhere in our society in recent years, which, however, has not yet crystallized in a general consensus of cutting waste production in our daily lives. Many methods could achieve sustainable development, methods that could not only improve waste management but could also lead to the production of industrially important chemicals, materials, and fuels, in essence, valuable end products from waste.

Waste valorization is the process of converting waste materials into more useful products including chemicals, materials, and fuels. Such concept has already existed for a long time, mostly related to waste management, but it has been brought back to our society with renewed interest due to the fast depletion of natural and primary resources, the increased waste generation and landfilling worldwide and the need for more sustainable and cost-efficient waste management protocols. Various valorization techniques are currently showing promise in meeting industrial demands. One among such promising waste valorization strategies is the application of flow chemical technology to process waste to valuable products. A recent review of Ruiz et al. [1] highlighted various advantages of continuous flow processes particularly for biomass and/or food waste valorization which included reaction control, ease of scale-up, efficient reaction cycles producing more yield, and no required catalyst separation. Although flow chemistry has been known to be used in industries for other processing methodologies, it still remains to be used in biomass/waste valorization—a limitation caused by the large energy needed to degrade highly stable biopolymers and recalcitrant compounds (e.g., lignin). The deconstruction of such biopolymers, most of the time, requires extreme conditions of pressure and temperature—conditions achieved by microwave heating, which is another green valorization technology. These requirements are not simple to satisfy and various techniques (e.g., microwave irradiation) need to be combined to satisfy the prerequisites for a successful transformation of waste. However, the main challenge for this combination is on the scale-up itself. As conceptualized by Glasnov et al. [2] microwave and flow chemistries maybe coupled by attaching back-pressure regulators to flow devices. This approach can revolutionize industrial valorization since it will synthesize products fast (due to microwave heating) on one continuous run (flow process). Although the approach presented is possible, the main challenge of temperature transfer from microwave to flow remains to be solved. A buildup of temperature gradient inside the instrument could lead to various instrument inefficiencies.

Another valorization strategy is related to the use of pyrolysis in the synthesis of fuels. This involves biomass heating at high temperatures in the absence of oxygen to produce decomposed products [3, 4]. Although pyrolysis is a rather old method for char generation, it has been recently utilized to produce usable smaller molecules from very stable biopolymers. This method has been particularly employed in the production of Bio-Oil (a liquid, of relatively low viscosity that is a complex mixture of short-chain aldehydes, ketones, and carboxylic acids). In a study by Heo et al. [5], several conditions for the fast pyrolysis of waste furniture sawdust were studied, and it was found that bio-oil yields do not necessarily increase with temperature. The optimized pyrolysis temperature was set at 450°C (57% bio-oil yield) using a fluidized bed reactor. The reason for the nonlinear dependence bio-oil yield/temperature is the possible decomposition of small molecules into simpler gases. This theory is supported by the increase in the amount of gaseous products found at increasing temperatures. A separate study by Cho et al. [6] employed fast pyrolysis under a fluidized bed reactor to recover BTEX compounds (benzene, toluene, ethylbenzene, and xylenes) from mixed plastics. The highest BTEX yield was obtained at 719°C. The pyrolysis of cotton stalks was also reported to produce second generation biofuels [7]. The study found that at much higher temperatures of pyrolysis the amounts of H_2 and CO collected increased, while CO_2 levels lowered. The decrease in CO_2 production could be due to the degradation of the gas at much higher temperatures producing CO and O_2. More recently, synergy between these first proposed technologies (microwave and pyrolysis) has been also reported to constitute a step forward toward more environmentally friendly low temperature pyrolysis protocols for bio-oil and syngas production [8]. Microwave-assisted pyrolysis of a range of waste feedstock can provide a tuneable and highly versatile option to syngas with tuneable H_2/CO ratios or bio-oil-derived biofuels via subsequent upgrading of the pyrolysis oil [8].

Aside from energy applications, pyrolysis can also be used to produce advanced materials including carbon nanotubes and graphene-like materials, which have a wide range of applications. These studies along with many others in literature illustrate the potential of pyrolysis to convert waste materials into valuable chemicals.

A third green method of valorization would be on the use of biological microorganisms to degrade complex wastes and produce fuel. The method is used by taking advantage of cellulose (or any biopolymer) degrading enzymes by microorganisms as demonstrated by Wulff et al. [9]. In their work, cellulase Xf818 was isolated from the plant pathogen *Xylella fastidiosa* (known to cause citrus variegated chlorosis in plants). The gene responsible for the enzyme was also probed and then later on expressed on *Escherichia coli*. Such enzyme was found to be mostly active in the hydrolysis of carboxymethyl celluloses, oat spelt xylans, and wood xylans.

Bioconversion has been under intensive research for the past years, and one of the most significant advances in the field relates to the possibility of a synthetic control of microorganisms' metabolic pathways to produce favorable metabolic processes, which will in turn increase the yield of products. A notable example is the use of a bioengineered *E. coli* to produce higher alcohols including isobutanol, 1-butanol, 2-methyl-1-butanol, 3-methyl-1-butanol and 2-phenylethanol from glucose [10]. The protocol was amenable for the conversion of 2-ketoacid intermediates (from amino acid biosynthesis) into alcohols by amplifying expression of 2-ketoacid decarboxylases and alcohol dehydrogenases.

To model the design for isobutanol, the gene *ilv*IHCD was over-expressed with the $P_L lacO_1$ promoter in a plasmid to amplify 2-ketoisovalerate biosynthesis. Other genes were tested such as alsS gene (from *Bacillus subtilis*) to further improve the alcohol yield while some genes responsible for by-product formation (*adhE, ldhA, frdAB, fnr, pta*) and pyruvate competition (*pflB*) were silenced. Overall, the isobutanol yield reached ~300 mmol/L (22 g/L) under microaerobic conditions.

The three presented strategies (microwave, pyrolysis, and bioengineering) represent some of the most important valorization methodologies. With the rapid advancement of these fields in waste valorization, it is expected that most industrial sustainability practices will have a different focus in various future scenarios.

Waste valorization is currently geared toward three sustainable paths: one would be on the production of fuel and energy to replace common fossil fuel sources and in parallel on the production of high-value platform chemicals as well as useful materials. Fossil-based fuels are clearly diminishing in supply and this has caused a global environmental concern due to rapidly rising emissions of fossil fuel by-products (both for processing and actual use). Because of this, waste valorization for energy and fuels are not only geared toward a sustainable fuel source but also toward a more benign fuel fit for an industrial up-scale. According to the Netherlands Environmental Assessment Agency, global CO_2 emissions reached an all-time high in 2011 at around 34 billion tonnes of greenhouse gases (GHGs) [3]. Close to 90% of these emissions derive from fossil fuel combustion. Other toxic gases such as volatile organic compounds, nitrogen oxides (precursors of toxic ozone) and particulates come together with GHGs. In a more than likely scenario of a minimum of 2.5% energy demand growth per year, it is necessary to substitute fossil fuels progressively with cleaner fuel sources. Biomass combustion for electricity and heat production was reported to be less costly, providing at the same time a larger CO_2-reduction potential [11]. Many studies also have shown convincing proof that the use of biomass for energy applications could be a highly interesting solution and cleaner technology for the future [12-14].

Another direction of waste valorization aims to produce high-value chemicals from residues including succinic acid (SA) [15], furfural and furans [16], phenolic compounds [17], and bioplastics [18]. These can be produced via chemical, chemo-enzymatic, and biotechnological approaches (e.g., solid state fermentation) but depending on the type of residue some compounds (e.g., essential oils, chemicals, etc.) can even be produced upon extraction and isolation [19]. The production of biomass-derived chemicals is a sustainable approach since it maximizes the use of resources and, at the same time, minimizes waste generation.

The major strength of biotechnology is its multidisciplinary nature and the broad range of scientific approaches that it encompasses. Among the broad range of technologies with the potential to reach the goal of sustainability, biotechnology could take an important place, especially in the fields of food production, renewable raw materials and energy, pollution prevention, and bioremediation. At present, the major application of biotechnology used in the environmental protection is to utilize microorganisms to control environmental contamination. Developing biotechnology could be a solution for these problems—this will also be given emphasis in this review.

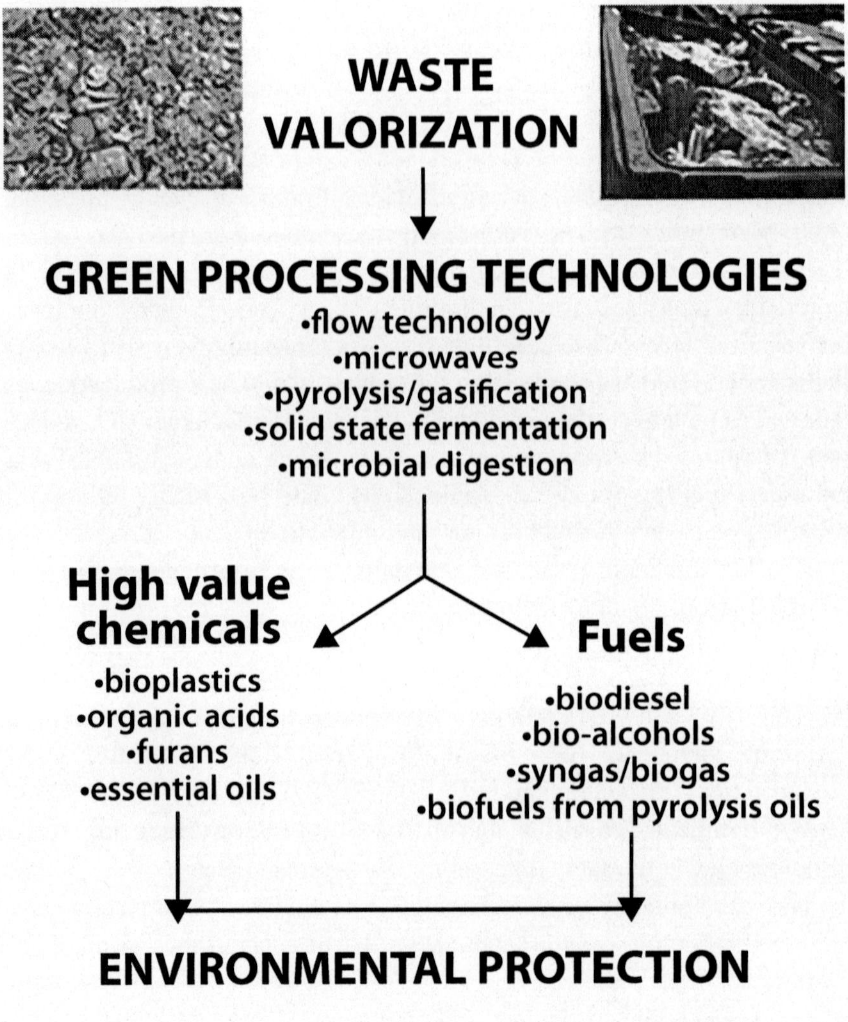

FIGURE 1: Valorization is essentially a concept of recycling waste into more usable industrial chemicals. Using established Green Processing technologies, various types of waste can be converted into high-value chemicals and fuels with the purpose of minimizing waste disposal volumes and eventually protecting the environment.

Although waste valorization is an attractive approach for sustainability, on a large scale perspective, the purification, processing, and even the degradation of stable natural polymers (e.g., lignin) into simple usable chemicals still remain a significant challenge (Fig. 1).

In recent years, there have been increasing concerns in the disposal of food waste. The amount of food waste generated globally accounts for a staggering 1.3 billion tonnes per year. Apart from causing the loss of a potentially valuable food source or the regenerated resource, there are problems associated with the disposal of food waste into landfills. With this imminent waste management issue, food waste should be diverted from landfills to other processing facilities in the foreseeable future. In Hong Kong, there are 3600 tonnes of food waste generated (Table 1), 40% of which is made up of municipal solid waste (MSW). Fifty two percent (52%) of the MSW generated is dumped into landfills [20]. It is estimated that by 2018, all current landfill sites in Hong Kong will be exhausted.

TABLE 1: Composition of waste in Hong Kong [20]

Waste	Tonnes/day
Municipal solid waste	9000
Domestic waste (including food waste)	6000 (2550)
Commercial and industrial waste (including food waste)	3000 (1050)
Construction waste	3350
Sewage sludge	950
Other waste	200
Total	13,500

Although the problem of food waste is commonly found over the world, the systems of food waste processing can only be formulated at the local community level with the consideration of the area-specific characteristics. These include regional characteristics and composition of waste, land availability, people's attitude and so forth. However, due to the lack of the local study concerning the suitability of food waste processing technologies for Hong Kong, this review is important to provide a few suggestions for the authorities to contemplate the adoption of a strategy on food waste disposal.

This contribution has been conceived to provide an overview on recent development of waste valorization strategies (with a particular emphasis on food waste) for the sustainable production of chemicals, materials, and fuels, highlighting key examples from recent research conducted by our groups. Reports on the development of green production strategies from waste and key insights into the recent legislation on management of wastes worldwide will also be discussed. The incorporation of these processes in future biorefineries for the production of value-added products and fuels will be an important contribution toward the world's highest priority target of sustainable development.

2.2 WASTE: PROBLEMS AND OPPORTUNITIES

In recent years, problems associated with the disposal of food waste to landfills lead to increased interest in searching for innovative alternatives due to the high proportion of organic matter in food waste. First generation food waste processing technologies include waste to energy (e.g., anaerobic digestion), composting, and animal feed. Based on the characteristics of food waste, an integrated approach should be adopted with the focus on food waste reduction and separation, recycling commercial and industrial food waste, volume reduction of domestic food waste and energy recovery from food waste.

2.2.1 SOURCES, CHARACTERIZATION, AND COMPOSITION OF WASTE

The large amounts of waste generated globally present an attractive sustainable source for industrially important chemicals. Food waste including garbage, swill, and kitchen refuse [21], can be generally described as any by-product or waste product from the production, processing, distribution, and consumption of food [22].

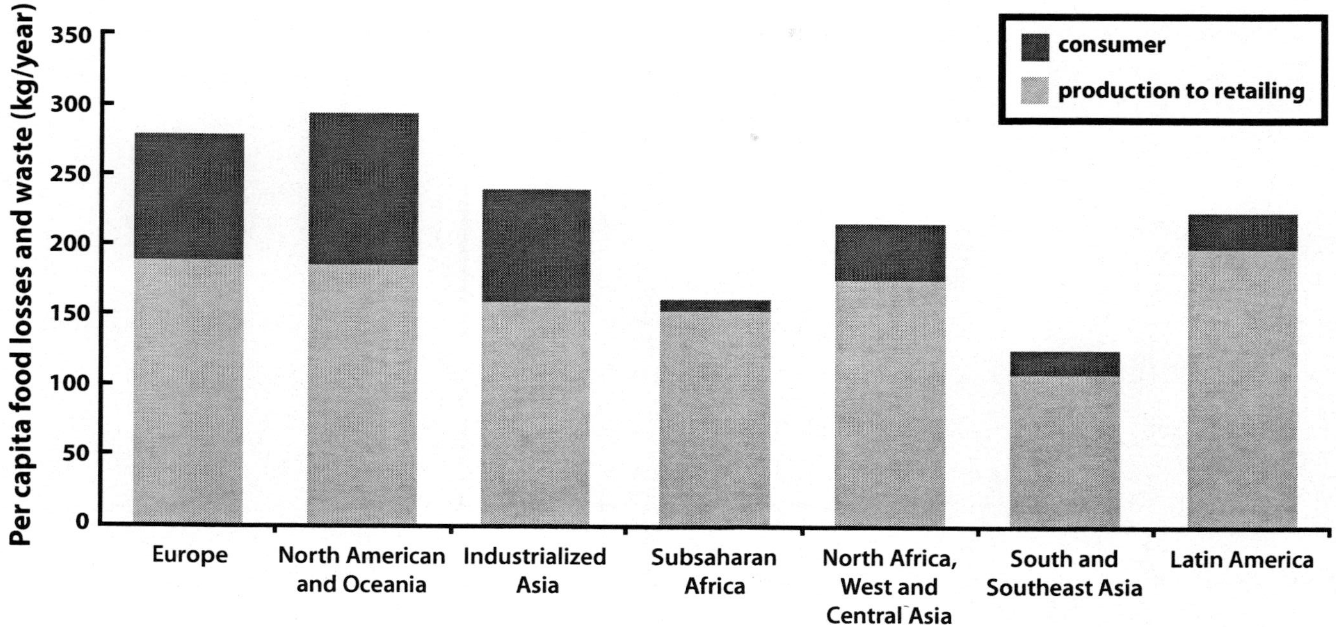

FIGURE 2: Per capita food losses and waste, at consumption and pre-consumption stages, in different regions [19].

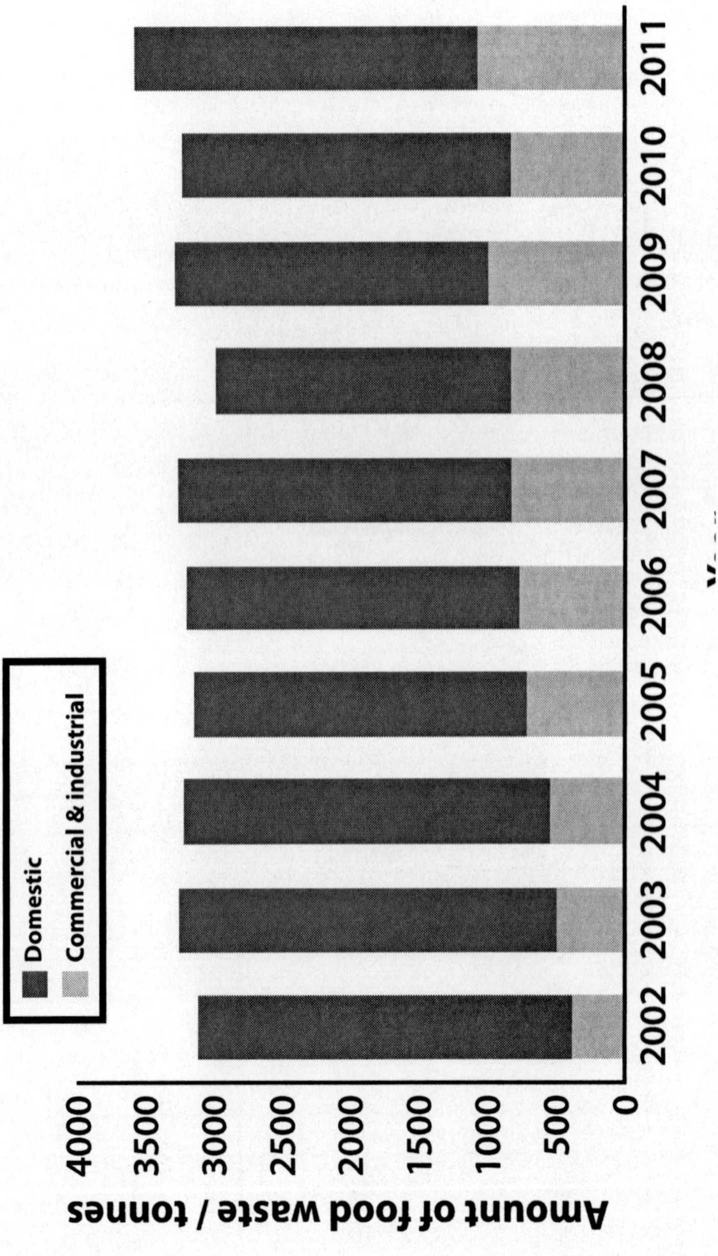

FIGURE 3: The amount of food waste generated daily in Hong Kong from 2002 to 2011 [25].

TABLE 2: Characteristics of reported domestic food waste [18]

Source	Characteristics			Country
	Moisture content (%)	Volatile solid/ total solid (%)	Carbon/ nitrogen	
A dining hall	80	95	14.7	Korea
University's cafeteria	80	94	NA[a]	Korea
A dining hall	93	94	18.3	Korea
A dining hall	84	96	NA	Korea
Mixed municipal sources	90	80	NA	Germany
Mixed municipal sources	74	90–97	NA	Australia
Emanating from fruit and vegetable, markets, household and juices centers	85	89	36.4	India

[a]*NA, not available.*

The definition of food waste is, however, different in different countries or cities. In the European Union, food waste is defined as "any food substance, raw or cooked, which is discarded, or intended or required to be discarded." The United States Environmental Protection Agency (EPA), on the other hand, defines food waste as "Uneaten food and food preparation waste from residences and commercial establishments such as grocery stores, restaurants, and produce stands, institutional cafeterias and kitchens, and industrial sources including employee lunchrooms." In the United Nations, "Food waste" and "Food loss" are distinguished. Food losses refer to the decrease in food quantity or quality, which makes it unsuitable for human consumption [23] while food waste refer to food losses at the end of the food chain due to retailers' and consumers' behavior [24]. All in all, food waste includes not just wasted foodstuffs, but also uncooked raw materials or edible materials from groceries and wet market.

Food waste is generally characterized by a high diversity and variability, a high proportion of organic matter, and high moisture content. Table 1 summarizes some reported characteristics of food waste, indicating moisture content of 74–90%, volatile solids to total solids ratio (VS/TS) of 80–97%, and carbon to nitrogen ratio (C/N) of 14.7–36.4 [25]. Due to these properties, food waste disposal constitutes a significant problem due to the growth of pathogens and rapid autoxidation [26]. As there are already many different microorganisms in food waste, the high rate of microbial activity and the amount of nutrients in food wastes facilitate the growth of pathogens, which cause the concern for foul odor, sanitation problems, and could even lead to infectious diseases. The high moisture contents [23] also increase the cost of food waste transportation. Food waste with high lipid content is also susceptible to rapid oxidation. The release of foul-smelling fatty acids also adds difficulties to the storage of treatment of food waste (Table 2).

According to a study commissioned by the United Nations Food and Agriculture Organization (UNFAO) in 2011, 1.3 billion tonnes of food waste is generated per year and roughly one third of food produced for human consumption is lost or wasted globally. The report also noted that food waste of industrialized countries and developing countries have different characteristics. Firstly, increasingly important quantities of food waste are generated in industrialized countries as compared to volumes observed in developing countries on a per capita basis. Figure 2 shows the per capita food loss in Europe and North America is 280–300 kg/year. In contrast, the food loss per capita in sub-Saharan Africa and South/Southeast Asia accounts for 120–170 kg/year. Also, food waste is mainly generated at retail and consumer levels in industrialized countries. Comparatively, food waste is generated in developing countries mainly at postharvest and processing levels, supported by the per capita that is, food waste generated by consumer levels in Asia is only 6–11 kg/year [27].

Interestingly, the amount of food waste generated for example in Hong Kong is staggering. Figure 3 shows an increasing trend of the food waste generated daily from 3155 tonnes in 2002 to 3484 tonnes in 2011. Although the disposal of food waste in landfills was found to be the most economical option [28], it causes numerous problems in landfill sites. As landfilling disposal generally buries and compacts waste under the ground, the decomposition of food waste produces methane, a GHG that is twenty-one times powerful than carbon dioxide (CO_2) under anaerobic environment conditions. Such production can in fact remarkably affect the environment in the area as some reports indicated that around 30% of GHG produced in Hong Kong are generated in landfill sites [29]. Methane is also flammable and may lead to fires and explosions upon accumulation at certain concentrations. In addition, the decomposition of food waste develop unpleasant odor as well as leachates and organic salts that could damaging landfill liners, leaching out heavy metals and resulting in contamination of ground waters [30].

Valorization research has evolved through the years, with many techniques and developments achieved in recent decades. Waste feedstock including bread, wheat, orange peel residues, lignocellulosic sources, etc. are currently explored as sources of chemicals and fuels. On a recent review by Pfaltzgraff et al. [31], it was noted that the valorization of food wastes into fine chemicals is a more profitable and less energy consuming as compared to its possibilities for fuels production. Because of this, related waste processing technologies, particularly related to the production of fuels, have also been proposed to address energy efficiency and profitability from a range of different feedstocks. Toledano et al. [32, 33] reported a lignin deconstruction approach using a novel Ni-based heterogeneous catalyst under microwave irradiation. Different hydrogen donating solvents were explored for lignin depolymerization, finding formic acid as most effective hydrogen donating reagent due to the efficient generation of hydrogen for hydrogenolysis reactions (from its decomposition into CO, CO_2, and H_2) and its inherent acidic character that induces acidolytic cleavage of C-C bonds in lignin at the same time. The heterogeneous acidic support also acted as a Lewis acid, coordinating to lignin thereby promoting acidic protonation, and eventually dealkylation and deacylation reactions (Fig. 4). Figure 5 shows the structural complexity of the lignin biopolymer. Lignin deconstruction to simple aromatics including syringaldehyde, mesitol, and related compounds could serve the basis for a new generation of renewable gasolines [34].

Simple phenolic compounds with potential antioxidant properties can also be derived from cauliflower by-products [35]. The proposed valorization strategy comprised a combined solvent-extraction step using an organic solvent together with a polystyrene resin (Amberlite XAD – 2) to recover most phenolics prior to high performance liquid chromatography (HPLC) analysis. Kaempferol-3-O-sophoroside-7-O-glucoside and its sinapoyl derivative kaempferol-3-O-(sinapoylsophoroside)-7-O-glucoside were obtained as main extracted components. A separate study by Sáiz et al. [36] also proved near-infrared (NIR) spectroscopy was a highly useful technique to characterization online of alcohol fermentation from onions. Along with multivariate calibration, this technique can lead to the analysis of samples with complex matrices without a prior sample preparation. One approach to a greener characterization method would be the coupling of a chromatographic technique to a flow instrument. This coupling has been shown to work in studies on metal analysis [37], online derivatization and separation of aspartic acid enantiomers [38], as well as for an enzyme inhibition assay [39], but it has not yet been shown to be successful for waste valorization.

2.2.2 DEVELOPMENT OF GREENER VALORIZATION STRATEGIES

There are numerous options for waste processing and/or recycling in the world. Composting, regenerated animal feed and bedding, incineration, anaerobic digestion, and related first generation strategies have been proposed and investigated for a long time. Some of these techniques have been successful in making their way to commercialization. Considering the storage problem and the large amount of food waste generated every day, food waste processing facilities have to be in a mega-scale size with enough treatment capacity to handle numerous tonnes of food wastes daily. It definitely requires a large initial investment for setting up the industrial scale facilities. Also, in case of off-site processing, the large volume and great weight of food waste adds difficulty since the collection of food waste significantly increases the transportation cost and time. Besides, the variation in composition of food wastes, affects the quality of regenerated products, such as compost and animal feed. Therefore, it decreases the product's competiveness in the market.

As demonstrated in the above-mentioned examples, valorization may be carried out under different conditions depending on the target components needed. Before reaching an industrial upscale, enhancement of valorization product yield may be done by careful variation of the valorization strategies, in particular advanced protocols able to diversify on feedstock and end products obtained from them. Currently, an active area of research relates to catalytic valorization strategies using solid acid catalysts [41, 42]. One example of a green protocol on valorization of waste oils to biodiesel was provided by Fu et al. [43], in which a superacid was prepared by adding a sulfuric acid solution to zirconium hydroxide powder. Under optimum reaction conditions, 9:1 MeOH/oil molar ratio, 3% (w/w) catalyst, and 4 h reaction time at 120°C, biodiesel yield reached 93.2%. Apart from metal supports functionalized with acids, carbon-based catalysts for waste valorization are also attractively developed protocols. Aside from being an easily separable reaction component, functionalized carbonaceous materials can also be recyclable. In a study by Clark et al. [44], carbonaceous materials from porous starches (Starbons®, Department of Chemistry, University of York, York, UK) functionalized with sulfonated groups were found to have a catalytic activity 2–10 times greater to those of common microporous carbonaceous catalysts in a range of chemistries including biodiesel production from waste oils and SA transformations in a fermentation broth. A separate study by Luque et al. [45] employed carbonaceous residues of biomass gasification as catalysts for biodiesel synthesis. The results showed good ester conversion yields from fatty acids to methyl esters. The abovementioned examples demonstrate that designer catalysts can be attractive options in the valorization of a range of waste feedstocks.

A promising sustainable approach would also be the use of ionic liquid-type compounds which can be derived from renewable feedstock such as the so-called deep-eutectic solvents [46-49] and even selected designer ionic liquids. These compounds are salts in their liquid states with very unique properties such as very low vapor pressure, thermal stability, and tunability based on different applications. A study by Ruiz et al. [50] presented a $-SO_3H$ functionalized Bronsted acid ionic liquids catalyzed synthesis of an important chemical precursor such as furfural from C5 sugars under microwave heating. Furfural yield varied from 40% to 85% depending on the type of Ionic liquid used and the feedstock employed in the process. It was shown that the ionic liquid 1-(4-Sulfonylbutyl)pyridinium tetrafluoroborate produced a yield of 95% for xylose conversion and 85% for furfural. Importantly, the protocol was amenable to the utilization of a biorefinery-derived syrup enriched in C5 oligomers, from which a 40–45% of furfural yield could be derived.

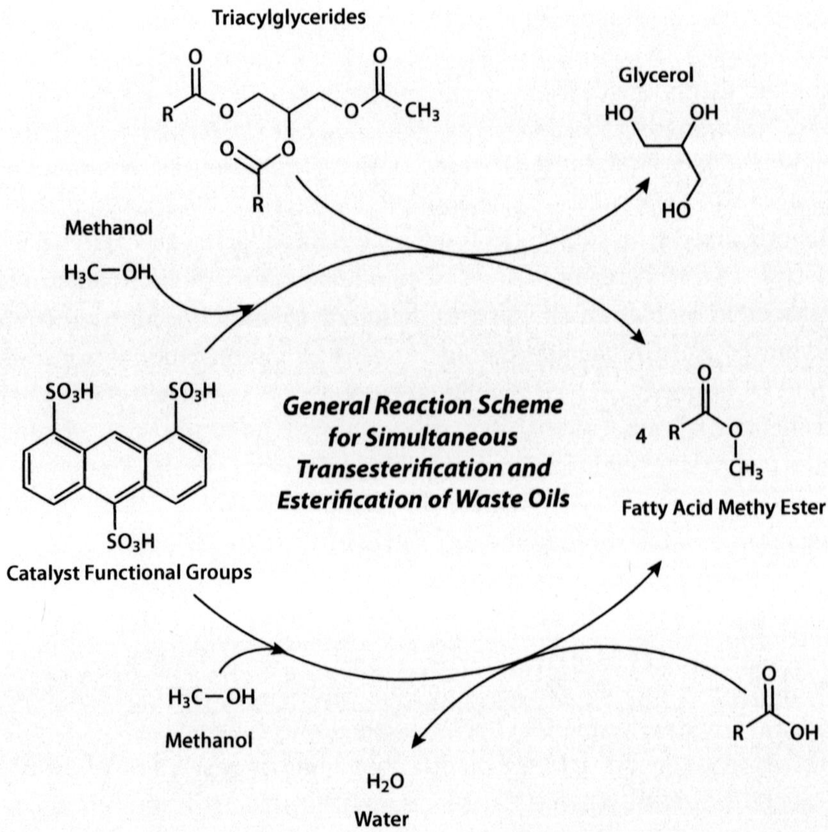

FIGURE 4: Simultaneous transesterification and esterification of waste oils using solid acid catalysts produced fatty acid methyl esters (a nonpolar component) along with water and glycerol (polar compounds) that separate out spontaneously from the reaction mixture forming two phases.

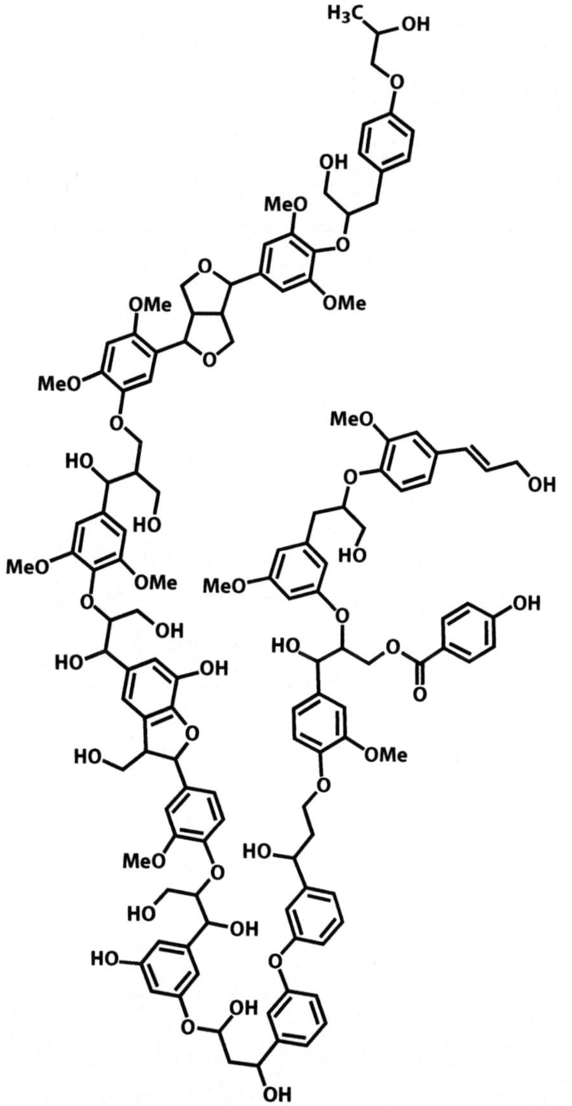

FIGURE 5: The structural complexicity of lignin being composed of aromatic compounds show its potential in different applications such as for fuel, and in the production of high-value chemicals (Image adapted from Stewart et al., [40]).

A separate study by Zhang et al. [51] showed that the direct conversion of monosaccharides, and polysaccharides to 5-hydroxymethylfurfural (5-HMF) may be accomplished using ionic liquids in the presence of Germanium (IV) chloride. Yields of the reaction could go as high as 92% depending on the reaction conditions used. The mechanism proposed by the researchers indicate the role of the $GeCl_4$ as a Lewis acid catalyst for the ring opening of the sugars, which is immediately followed by several dehydration steps to produce 5-HMF.

As alternative to these catalytic strategies, photocatalytic approaches to waste valorization could also serve the basis of innovative and highly attractive future valorization protocols. A recent review by Colmenares et al. [52] addressed the potential and opportunities of photocatalysis to convert lignin biomass into fine chemicals using designer TiO_2 nanocatalysts. These nanomaterials featuring doping agents (to lower the band gap of titania) have been shown to be effective in water splitting experiments (to form H_2 and O_2) to harness the potential of hydrogen as fuel. One of the earliest promising works of light-mediated degradation was shown by Stillings et al. [53] when they were able to degrade cellulose using Ultraviolet radiation. However, this has not been shown to be possible using visible light due to energy considerations. A photocatalytic approach to degradation may also be accomplished using functionalized graphenes (monolayers of sp2 carbon atoms in a honeycomb lattice known to have ballistic electron transport properties). Functionalized graphenes and composites with other semiconductors have been shown to exhibit degradation properties [54, 55], but this concept has not yet been applied to waste valorization strategies.

2.2.3 RECENT LEGISLATION ON WASTE MANAGEMENT

2.2.3.1 PHILIPPINES

In Metro Manila at Philippines, almost 3.5 kg of solid waste is generated per capita every day. This amount includes food/kitchen waste, papers, polyethylene terephthalate bottles, metals, and cans. Although most Metro Manila residents do not practice the open burning of waste, a necessary waste segregation is performed for ease of collection. Being the country's capital, and one of the world's most densely populated cities, Metro Manila generates over 2400 tons of waste everyday, which equates to a government spending of Php 3.4 billion (63 Million Euros) in collection and disposal. Not much legislation is available in the Philippines in terms of waste management. Although Republic Act 9003 (Solid Waste Management Act) has been passed last 2000, a recent 2008 study showed that it has not been properly implemented [56].

2.2.3.2 HONG KONG

One third of the food waste generated in Hong Kong come from the Commercial and Industry (C&I) sector, with the remaining percentage coming from households. In recent years, the amount of disposal of food waste from C&I sectors remarkably increased by 280% from 373 tonnes in 2002 to 1050 tonnes in 2011. It is anticipated that the food waste generated in Hong Kong will continue to rise, driven by the significant increase of the C&I food waste generation. The disposal of food waste (an organic waste which decomposes easily) to landfills is not sustainable, as it leads to rapid depletion of the limited landfill space. From the 2013 Policy Address by the Office of the Chief Executive in Hong Kong [57], there was a special emphasis on "Reduction of Food Waste" as stated in Section 142 below:

> *"Food waste imposes a heavy burden on our landfills as it accounts for about 40% of total waste disposed of in landfills. In addition, odour from food waste creates nuisance to nearby residents. The Government has recently launched the "Food Wise Hong Kong Campaign" to mobilise the public as well as the industrial and commercial sectors to reduce food waste. We will build modern facilities in phases for recovery of organic waste so that it can be converted into energy, compost and other products."* [57].

The Environmental Protection Department (EPD) has planned to develop Organic Waste Treatment Facilities (OWTF). Such facilities will adopt biological technologies—composting and anaerobic digestion to stabilize the organic waste and turn it into compost and biogas for recovery. The first phase of the OWTF will be constructed at Siu Ho Wan with a daily treatment capacity of 200 tonnes of source separated organic waste (Fig. 6). The second phase will be located at Sha Ling of North District with a daily treatment capacity of 300 tonnes of organic waste.

Waste reduction at source should be the top priority so as to reduce the amount of food waste generated. Successful examples for the implementation of MSW charging scheme in Asian cities such as Taipei, Taiwan, and Seoul, South Korea could effectively reduce the total amount of MSW by 50% in 10 years [20]. These governments introduced quantity/volume-based charging scheme to create financial incentives to change public's food waste-generating behavior to achieve waste reduction at source. In addition, they introduced prepaid designated food waste bag charging system so as to achieve source separation. Food waste together with plastic bags can undergo treatment without extra separation step in the treatment facilities.

2.3 WASTE VALORIZATION STRATEGIES: CASE STUDIES

Biological treatment technologies including anaerobic digestion and composting have been reported extensively in past years. Under anaerobic digestion, biogas is generated as main product. Takata et al. [58] reported the production of 223 m^3 biogas from 1 tonne of food waste. However, Bernstad et al. [59] reported that the yield of biogas production may vary depending on the composition of waste and the existence of detergent. Numerous studies show that the lack of enough nutrients limits the ability of enzymes to digest waste [60, 21]. This can divert waste from landfill, and thus prevent the emission of GHG to the environment. Also, the solid residues can be used as compost, which can reduce the amount of used chemical fertilizers. Economically, anaerobic digestion can generate electricity on-site and may reduce energy cost. Also, it can be adopted in sewage treatment facilities, thereby eliminating transportation costs. Another way to valorize waste is by incineration for energy recovery. However, burning food waste is an energy intensive process and may remove important functional groups from the treated feedstocks. The following sections report case studies of different feedstock in different countries to illustrate the potential of waste valorization for the production of materials, chemicals, and products.

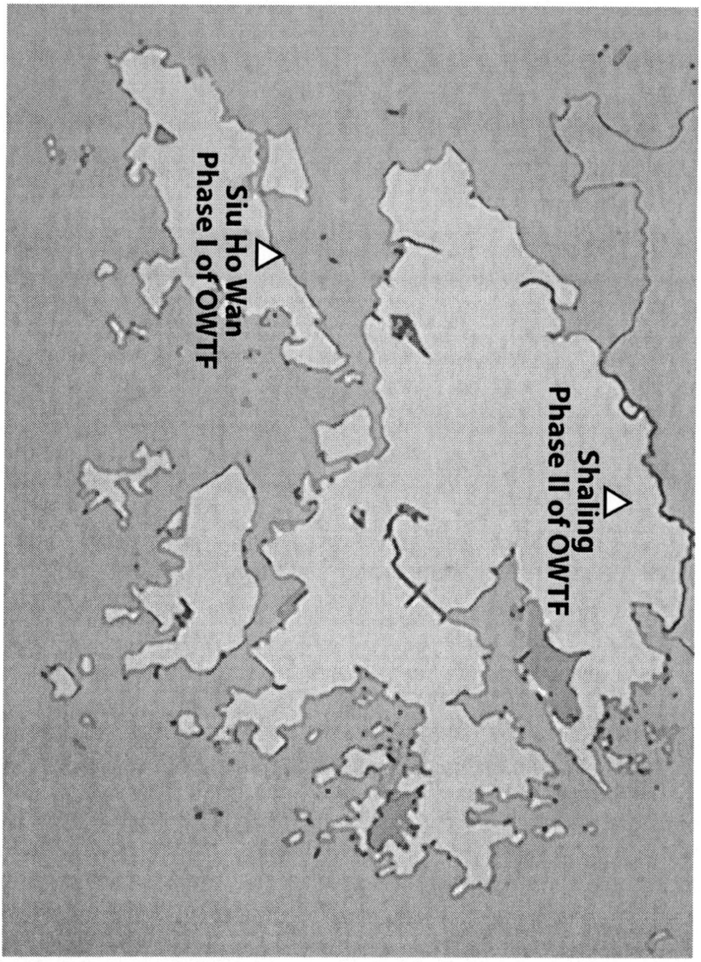

FIGURE 6: Map of Hong Kong indicates the location of the two organic waste treatment facility (OWTF) in Siu Ho Wan (Phase I) and Shaling (Phase II) [16].

2.3.1 UTILIZATION OF BAKERY WASTE IN THE BIOTECHNOLOGICAL PRODUCTION OF VALUE-ADDED PRODUCTS

Based on the large quantities of food waste generated at Hong Kong on a daily basis, Lin et al. have been recently focused on the valorization of unconsumed bakery products to valuable products via bio-processing in collaboration with retailer "Starbucks Hong Kong". Research was initially set on the production of bio-plastics poly(3-hydroxybutyrate) (PHB) and platform chemicals (e.g., SA) via enzymatic hydrolysis of non pretreated bakery waste, followed by fungal solid state fermentation to break down carbohydrates into simple sugars for subsequent SA or PHB fermentation. In the proposed biotechnological process, bakery waste serve as the nutrient source, including starch, fructose, free amino nitrogen (FAN), and trace amount of subsidiary nutrients. The nutrient content is listed in Table 3 below.

TABLE 3: Bakery waste composition (per 100 g) [61, 62]

Content	Pastry	Cake	Wheat bran
N/A, data not available.			
Moisture	34.5 g	45.0 g	N/A
Starch (dry basis)	44.6 g	12.6 g	N/A
Carbohydrate	33.5 g	62.0 g	15.0 g
Lipids	35.2 g	19.0 g	6 g
Sucrose	4.5 g	22.7 g	N/A
Fructose	2.3 g	11.9 g	N/A
Free sugar			1.5 g
Fiber	N/A	N/A	50 g
Protein (TN × 5.7) (dry basis)	7.1 g	17.0 g	14.0 g
Total phosphorus (dry basis)	1.7 g	1.5 g	N/A
Ash (dry basis)	2.5 g	1.6 g	N/A

In general, pastries have larger starch and lipid content to those of cakes; whereas cakes have higher sugar (fructose and sucrose) and protein content. Nevertheless, both types of bakery waste were proved to serve as excellent nutritional substrates for fermentative production of SA or bioplastics after hydrolysis. Our groups previously demonstrated that SA could be produced from wheat-based renewable feedstock [63-65] and bread waste [66] via fermentation. Similarly, production of biopolymers from various types of food industrial waste and agricultural crops was shown to be techno-economically feasible for replacing petroleum-derived plastics [67].

The key components in the project are illustrated in Figure 7. In the upstream processing, the bakery waste was collected from a Starbucks outlet in the Shatin New Town Plaza. A mixture of fungi comprising *Asperillus awamori* and *Asperillus oryzae* were utilized for the production of amylolytic and proteolytic enzymes, respectively. Macromolecules including starch and proteins contained in bakery waste were hydrolysed, expected to enrich the final solution in glucose and FAN. This hydrolysate was subsequently used as feedstock in a bioreaction by two different types of microorganisms (*Actinobacillus succinogenes* and *Halomonas boliviensis*) to produce (SA) and PHB, respectively.

Although food waste is a no-cost nutritional source, the application of commercial enzymes in upstream processing might not be cost-efficient. To reduce process costs, the degradation of bread and bakery waste has been previously studied [61, 66]. In these studies, *A. awamori* and *A. oryzae* were the fungal secretors of glucoamylase protease and phosphatase as well as a range of other hydrolytic enzymes that does not require any external addition of commercial enzymes.

According to Figure 8, glucose (54.2 g/L) and FAN concentrations (758.5 mg/L) were achieved at 30% (w/v) pastry waste after enzymatic hydrolysis. On the other hand, sucrose present in cake was hydrolyzed to form 1 mole of glucose and 1 mole of fructose. The glucose (35.6 g/L), fructose (23.1 g/L), and FAN concentrations (685.5 mg/L) were achieved at 30% (w/v) cake waste. Among all, waste bread hydrolysate contained the highest glucose and FAN concentrations, which were 104.8 g/L and 492.6 mg/L, respectively. These results clearly demonstrate the potential of utilizing bakery hydrolysate as generic feedstock for fermentations.

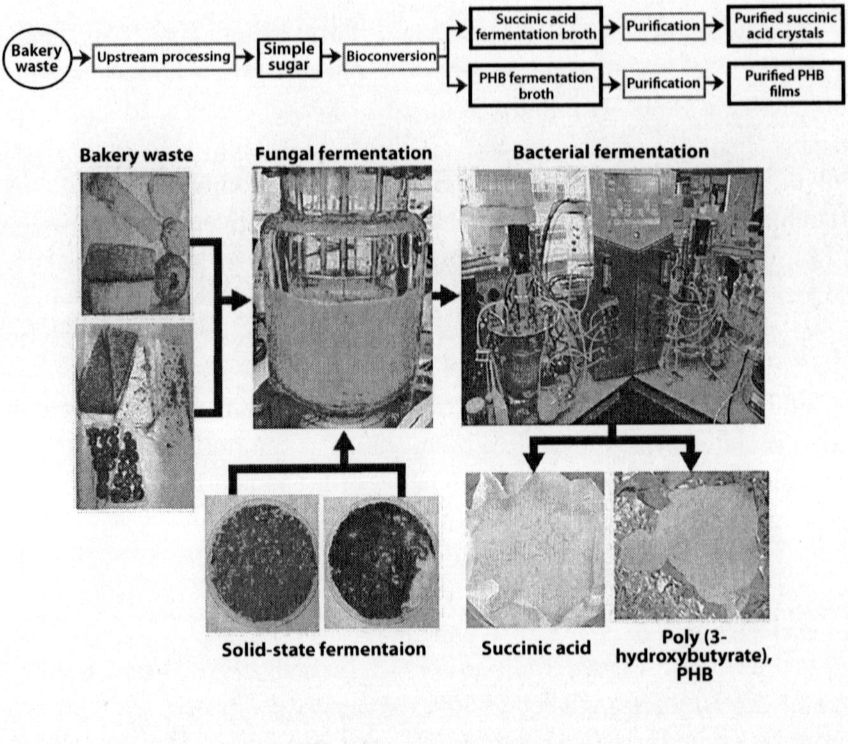

FIGURE 7: Flow chart of a bakery-based food waste biorefinery development, from bakery waste as raw material to succinic acid and poly(3-hydroxybutyrate), PHB as final products.

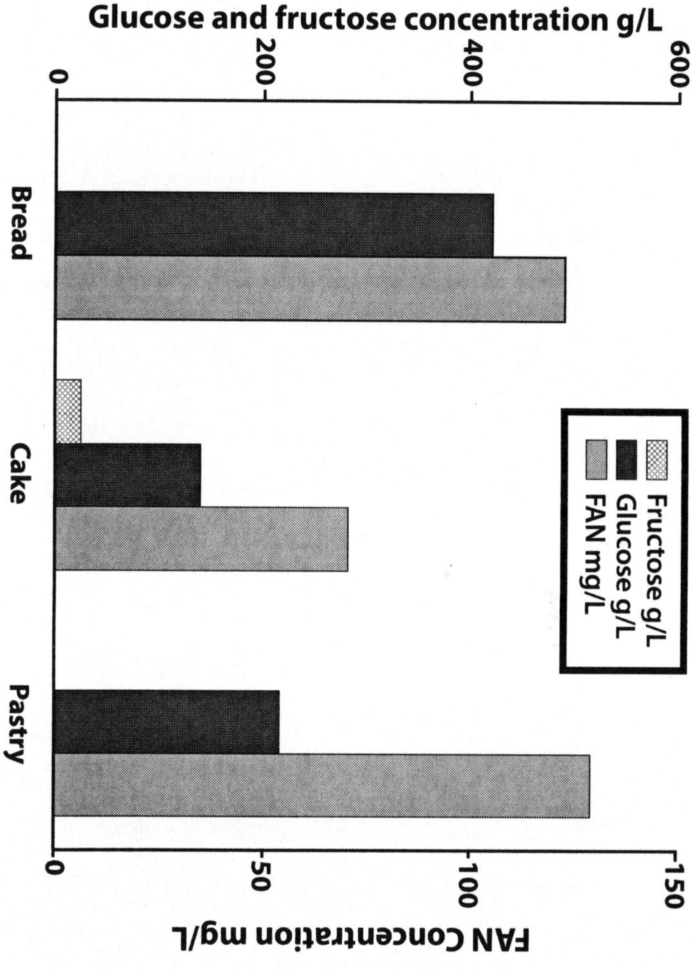

FIGURE 8: Sugars and FAN concentrations achieved from enzymatic hydrolysis using different bakery waste (30%, w/v) with *Aspergillus awamori* and *Aspergillus oryzae*.

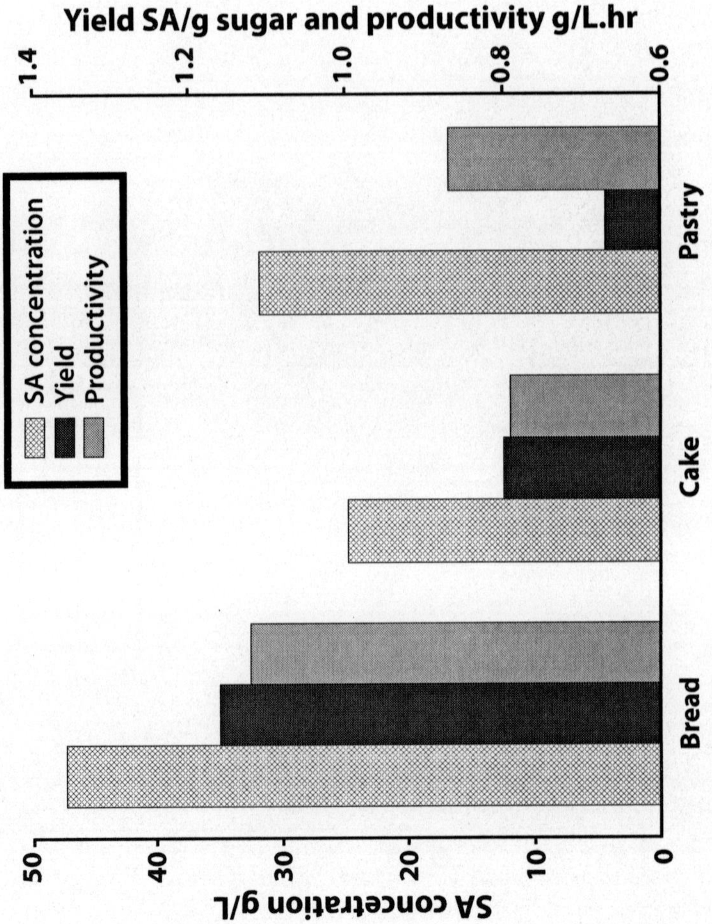

FIGURE 9: Succinic acid concentration, yield, and productivity in *Actinobacillus succinogenes* fermentations using different bakery hydrolysates.

TABLE 4: Comparison of succinic acid yields achieved using different food waste substrates with *Actinobacillus succinogenes*

Substrate	SA yield (g SA/g TS)	Overall SA yield (g SA/g substrate)	References
Wheat	0.40	0.40	[63]
Wheat flour milling by-product	1.02	0.087	[64]
Potatoes	N/A	N/A	[68]
Corncob	0.58	N/A	[69]
Rapeseed meala	0.115	N/A	[70]
Rapeseed mealb	N/A	N/A	[71]
Orange peel	0.58	Negligible	[72]
Bread	1.16	0.55	[66]
Cake	0.80	0.28	[61]
Pastry	0.67	0.35	[61]

N/A, not available. [a]Rapeseed meal is treated by diluted sulfuric acid hydrolysis and subsequent enzymatic hydrolysis of pectinase, celluclast, and viscozyme. [b]Rapeseed meal is treated by enzymatic hydrolysis using A. oryzae.

Batch fermentations on enzymatic hydrolysates were subsequently carried out to investigate the cell growth, glucose consumption as well as SA production. Cake hydrolysate consisting an initial sugar content of 23.1 g/L glucose and 18.5 g/L fructose, and pastry hydrolysate with an initial sugar content of 44.0 g/L glucose were both utilized as fermentation feedstock. At the end of fermentation, the remaining glucose was 5.2 g/L whereas fructose was 3.7 g/L. A final SA concentration of 24.8 g/L was obtained at the end point, which corresponded to a yield of 0.8 g SA/g total sugar and a productivity of 0.79 g/L.h (Fig. 9). The overall conversion of waste cake into SA was 0.28 g/g cake.

Compared with cake hydrolysates, pastry hydrolysates possessed larger concentrations of initial glucose (44.0 g/L). SA concentration continuously increased until sugar was depleted after 44 h. At the end of fermentation, the SA concentration reached 31.7 g/L, which corresponded to a yield of 0.67 g SA/g glucose and a productivity of 0.87 g/L.h.

TABLE 5: The overall performance of both defined medium and bakery hydrolysate fermentation for PHB production in terms of fermentation conditions and results

Batch no.	Fermentation medium	Fermentation mode	Feeding media	Fermentation time (h)	Glucose consumption (g)	CDW (g)	PHB production (g)	PHB content (%)
1	Defined (40 g/L glucose, 2 g/L yeast extract)	Batch	NIL	64.0	13.0	NIL	NIL	NIL
2	Defined (40 g/L glucose, 5 g/L yeast extract)	Batch	NIL	88.0	24.0	24.9	17.4	17.4
3	Defined (40 g/L glucose, 8 g/L yeast extract)	Batch	NIL	75.0	59.9	9.2	4.3	4.3
4	Pastry hydrolysate	Batch	NIL	23.5	32.8	NIL	NIL	NIL
5	Pastry hydrolysate	Fed-batch	Glucose solution	135.5	112.3	5.7	2.1	2.1
6	Pastry hydrolysate	Fed-batch	Pastry hydrolysate	67.0	208.8	38.2	3.6	3.6
7	Pastry hydrolysate	Fed-batch	Pastry hydrolysate	87.0	359.9	15.6	0.6	0.59
8	Cake hydrolysate	Fed-batch	Cake hydrolysate	63.0	200.5	11.6	2.9	2.9

CDW, cell dry weight.

SA production achieved from various food waste residues has been compared in Table 4. It is clear that SA yields obtained when using cake and pastry wastes as feedstock were comparable or higher to those of other food waste-derived media.

2.3.2 BIOTECHNOLOGICAL PHB PRODUCTION USING BAKERY WASTE AND SEAWATER

Halomonas boliviensis has been utilized in fermentations for the bioconversion of bakery hydrolysate into PHB. This microorganism is a moderate halophilic and alkali tolerant bacterium that can produce PHB through fermentative processes under aerobic condition [73]. It was isolated from a Bolivian salt lake, and the rod-shaped *H. boliviensis* is able to survive and synthesize PHB under salty environment.

Table 5 shows a summary of PHB fermentation results using defined medium and bakery hydrolysates, namely cake and pastry hydrolysates. PHB yields suggested that a defined fermentation medium (40 g/L glucose, 5 g/L yeast extract) can provide an optimum PHB yield (72%). The lowest PHB yield (1–2%), as expected, was obtained under bakery hydrolysate fermentation media. This demonstrates that a defined medium with 40 g/L glucose and around 5 g/L yeast extract could provide sufficient nutrients for H. boliviensis to produce PHB efficiently.

The overall glucose consumption for defined medium fermentation in the batch mode ranged from 13 to 60 g. High initial nitrogen source could hinder PHB production by 10 times, as indicated from PHB yield obtained by defined medium. A similar effect could possibly lead to the low PHB yield observed in bakery hydrolysate fermentation. With the continuous supply of nitrogen source, *H. boliviensis* consumed glucose in a faster rate for PHB production, maintenance, and synthesis of other metabolites (six times higher overall glucose consumption with the feeding of bakery hydrolysate). *Halomonas boliviensis* synthesizes ectonie and hydroxyectonie as osmolytes as NaCl concentration increases in the cell's environment. Van-Thuoc et al. [74] reported the co-production of ectonie and PHB in a combined two-step fed-batch culture. Similarly, the formation of other primary metabolites such as ectoines in the bakery hydrolysate fermentation was observed in these studies. This consequently led to a lower PHB production when bakery hydrolysate and seawater were used as fermentation feedstocks. The highest overall yield of PHB production for the defined medium (with less glucose consumption and higher PHB production) was about 17% as compared to a rather low 3.5% observed for bakery hydrolysate.

In summary, this project is currently demonstrating the green credentials in the development of advanced food waste valorization practices to valuable products, which also include GHG reductions as well as the production of other air pollutants. Such a synergistic solution may be feasible for adoption by the Hong Kong Government as part of their strategy for tackling the food waste issue as well as for the environmentally friendly production of alternative platform chemicals and biodegradable plastics.

2.3.3 CHEMICAL VALORIZATION OF FOOD WASTE FOR BIOENERGY PRODUCTION

The valorization of waste to important chemicals can be accomplished through different approaches as discussed. Another potentially interesting approach to advanced valorization practices would be the chemical utilization of various waste raw materials for conversion into high-value products.

A case study of such integrated valorization is a recent study on the conversion of corncob residues into functional catalysts for the preparation of fatty acid methyl esters (FAME) from waste oils [75]. The design of the catalyst involved an incomplete carbonization step under air to partially degrade the lignin materials mostly present in corncobs, followed by subsequent functionalization via sulfonation to generate $-SO_3H$ acidic sites. The solid acid catalyst was then subjected to conditioning prior to its utilization in the conversion of waste cooking oil with a high content of free fatty acids (FFA) to biodiesel-like biofuels. The advantage of the designed solid acid catalyst, apart from being derived from food waste, is the possibility to conduct a simultaneous esterification of FFA present in the waste oil as well as transesterification of the remaining triglycerides also present in the oil (Fig. 4).

In this approach, the generation of two valuable products (a cheap solid acid catalyst and biodiesel-like biofuels) can be achieved starting from two food waste feedstock (corncobs and waste cooking oils). The solid acid catalysts were characterized using a range of techniques. Fourier transform infrared spectroscopy (FTIR) showed the presence of different functional groups including C=O, C-O, C-S, and aromatic C=C in the materials (Table 6). The catalytic activity of the solids also showed remarkable activity toward the conversion of waste cooking oils into biodiesel-like biofuels. A maximum of 98% yield to methyl esters could be obtained without prior purification of the oils. Importantly, kinetics of the transesterification reaction was significantly slower to those of the esterification of FFA present in the oil. Despite a low $-SO_3H$ loading (1 wt% S, 0.16 mmol/g $-SO_3H$), the catalytic activity was still high, indicating a possibly different surface functionality (Fig. 10).

TABLE 6: Summary of bands observed in IR analysis for all sulfonated samples

Frequency of band	Corresponding functional group
1700 cm^{-1}	C=O
1597 cm^{-1}	C=C aromatic
1219 cm^{-1}	S=O
1029 cm^{-1}	C-S

The recyclability of the solid acid catalysts still, however, needs to be further optimized. Catalysts were found to deactivate quickly (after two uses) due to the aqueous promoted decomposition and hydrolysis of the sulfonated groups in the material [75]. Materials should be tested under different conditions of temperature, carbonizing atmosphere, and even pressure to improve the stability and robustness of the catalyst for the selected process. Nevertheless, this study provides a promising proof of concept of the potential of an integrated valorization of various waste raw material into valuable end products and biofuels as it avoids the pretreatment of the waste oils (generally required to reduce the high FFA content to allow the conventional base-transesterification process to take place avoiding the formation of undesirable soaps and emulsions) and generates a relatively pure biofuel from a residue using an environmentally friendly and cheap solid acid catalyst.

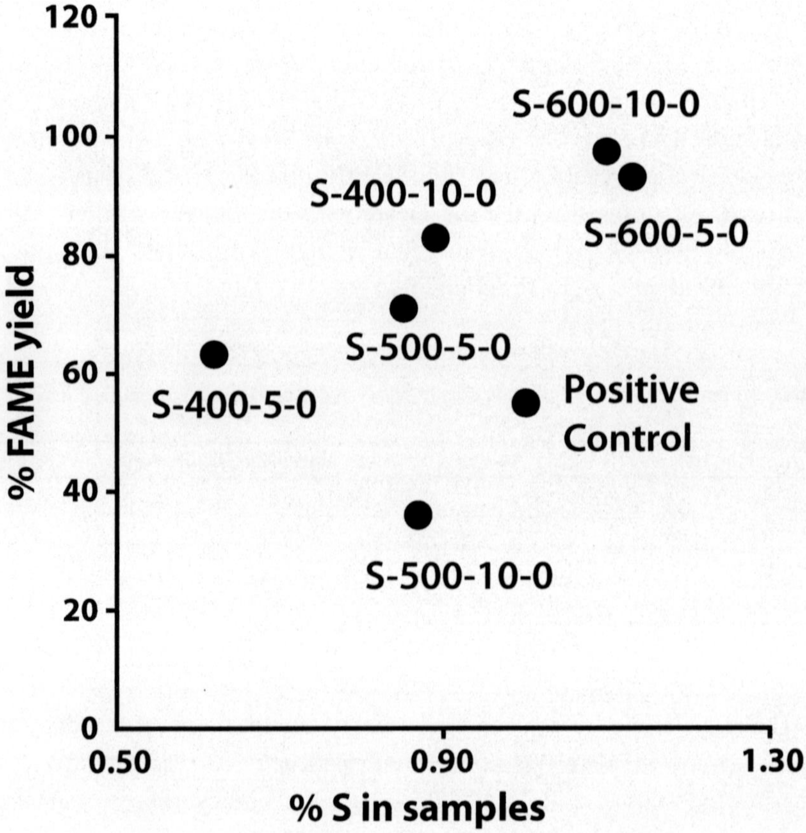

FIGURE 10: Plot of the FAME yield of the samples versus the %S content (A, material carbonized at 400°C for 5 h; B, carbonized at 400°C for 10 h; C, carbonized at 500°C for 5 h; D, carbonized at 500°C for 10 h; E, carbonized at 600°C for 5 h; F, carbonized at 600°C for 10 h). A higher degree of functionalization (higher %S) generally leads to improved FAME yields. Results also highlight the superior catalytic activities of sulfonated carbonaceous materials compared to blank (no conversion, data not shown) and the positive control referring to the homogeneously H_2SO_4 catalyzed reaction.

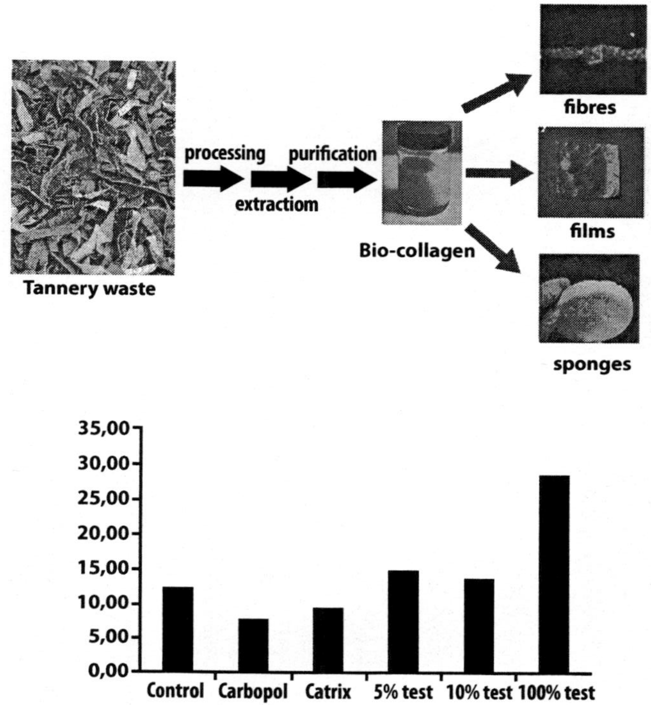

FIGURE 11: Meat and tannery-derived residues can be valorized to valuable collagenic biopolymers that can be formed into fibers, films, and sponges for various applications. Right plot depicts a comparison of activity in induced wounds in rats between pure and diluted collagen-extracted formulations (100%, 5% and 10% test, respectively) and a control sample (no treatment) and commercial formulations (Carbopol, Catrix).

2.3.4 TAILORED-MADE HEALING BIOPOLYMERS FROM THE MEAT INDUSTRY

The meat industry generates enormous quantities of solid waste [76]. Managing such residues entails a significant problem for the sector as many of the generated by-products and residues are prone to degradation and microbial contamination. However, an important part of these residues are rich in various added value products, which upon extraction could constitute a source of interesting revenues for these industries. Among the most promising compounds from meat industry-derived by products, we can include oily fats and collagen [77]. Valorization of the aforementioned waste fats from slaughterhouses and meat processing industries to biodiesel-like biofuels has been studied via esterification/transesterification using different types of catalysts and protocols, which entailed in some cases a pretreatment and refining of the fat [78]. In principle, these will be, however, conducted in a similar way to that reported in the previously showcased study of waste cooking oils valorization.

Comparatively, collagen-containing residues (e.g., bovine hides) are increasingly important residues from the meat industry that are often derived to leather processing companies. Interestingly, the significant amounts of collagen present in such samples are not that well known [77].

Collagen is a ubiquitous and most relevant biopolymer in vertebrates [77-79], which possesses a highly interesting versatility to be employed in a wide range of applications in different areas from regenerative medicine to cosmetics and veterinary. Extraction and stabilization of collagenic biopolymers from waste, particularly related to their physical properties, (e.g., via cross-linking) constitutes an innovative pathway toward the production of novel potentially industrial products (e.g., tissue engineering, wound healing, antimicrobial aposits, etc.). Cross-linking methodologies can in principle generate additional bond formation to stabilize polymers with additional benefits on physical properties including swelling and flexibility.

In the light of these premises, a recent example on the extraction, cross-linking and purification of collagenic biopolymers from splits (pickled hides) and the so-called wet-white hides (from tannery-derived hides treated with glutaraldehyde or phenolic compounds for chromium-free leather production) demonstrated the possibility to obtain valuable end products based on tailored-made biocollagen with improved mechanical properties, stabilized structures, and desired molecular weight ranges, which could be employed in wound healing acceleration in rats [80, 81]. Interestingly, these biopolymers could be easily shaped into various forms including fibers, sponges, and/or films, paving the way to the development of potentially novel biomaterials for different biomedical applications (Fig. 11). A simple hydrolytic process was able to extract the collagen, followed by subsequent cross-linking to stable biopolymers or direct application upon purification by ultrafiltration as unguent for induced wounds in rats [80, 81].

Maximum yields of biopolymer extracted were obtained at 0.25 mm grinding size and the use of diluted acetic acid as hydrolytic agent (24 h, room temperature), for which also a minimum swelling (better biopolymer properties) was also observed. Interestingly, samples obtained from splits exhibited a better desirability to those extracted from wet-white hides. Isolated biopolymers possessed molecular weight of ca. 300 kDa, in contrast to conventional collagen derivatives for which molecular weights are usually within 15–50 kDa for hydrolysates [82] and 50–200 kDa for gelatine [83].

Biocollagenic materials were found to be very attractive and highly useful in treatment of induced burns/wounds in mice (Fig. 11), showing in all cases improved tissue regeneration and wound healing as compared to untreated wounds and commercial formulations including Carbopol and Catrix. Even diluted formulations containing 5 wt% of the collagenic biopolymer (Fig. 11, right plot) were found to provide improved results.

2.4 FUTURE PROSPECTS AND CONCLUSIONS

Excessive disposal of food and plastic waste are deteriorating the landfill issue in many parts of the World. Waste valorization is an attractive concept that has gained increasing popularity in many countries nowadays due to the rapid increase in generation of such waste residues. Because of this, researchers are not only developing valorization strategies but also focusing on the design of greener materials utilizing a range of green technologies. One example of this could be the synthesis of magnetically separable substances [84–87]. Not only are these able to catalyze the necessary conversion, they are also economically attractive due to their simple preparation [84, 55, 85]. Also, the production of carbon-based catalysts maybe continued for research, but greener preparations (such as microwave-mediated functionalization) to lessen the energy investment, should be explored. Furthermore, the emergence of graphene as catalyst in many reactions should also be noted for valorization purposes.

As previously mentioned, an interesting valorization protocol to develop would be a photocatalytic approach. To accomplish such photocatalytic strategies, TiO_2, Pt/CdS/TiO_2 composite materials [88], TiO_2/$Ni(OH)_2$ [84] clusters may be used depending on the target, samples, and reaction conditions. Photodegradation has been shown to be possible toward many environmental pollutants such as chlorofluorocarbons [55], CO_2 [85], and NO [54], but whether these photoactive composites could degrade the stable polymeric structure of lignin/protein/carbohydrates is yet to be seen and perhaps understood. A recent study by Balu et al. [89], reports on the preparation of a TiO_2-guanidine-(Ni,Co)Fe_2O_4 photoactive material. The addition of the guanidine was made to lower the band gap of the material hence making it active under visible light. Testing the material to a model chemical reaction, using malic acid and the synthesized photomaterial produced simpler chemicals such as formic acid, acetic acid, and oxalic acid with a selectivity of around 80%. This study provides proof of concept that band gap engineering of semiconductors can lead to the development of photoactive materials that may be used selectively for waste valorization. A photocatalytic approach will most importantly address one of the major drawbacks of industrial valorization which is on the relatively large amounts of energy needed for processing and purification of products.

The conversion of a range of feedstock into valuable products including chemicals, biomaterials, and fuels has been demonstrated in three essentially different case studies to highlight the significant potential of advanced waste valorization strategies.

The incorporation of these and similar processes in future biorefineries for the production of value-added products and fuels will be an important contribution toward the world's highest priority target of sustainable development.

But perhaps the main and most important issue to be addressed for the sake of future generations, currently way overlooked, is society itself. The most extended perception of waste as a problem, as a residue, as something not valuable needs to give way to a general consensus of society in waste as a valuable resource. A resource, which obviously entails a significant complexity (from its inherent diversity and variability), but one that can provide at the same time an infinite number of innovative solutions and alternatives to end products through advanced valorization strategies. These will need joint efforts from a range of disciplines from engineering to (bio)chemistry, bio(techno)logy, environmental sciences, legislation, and economics to come up with innovative alternatives that we hope to see leading the way toward a more sustainable bio-based society and economy.

REFERENCES

1. Serrano-Ruiz, J. C., R. Luque, J. M. Campelo, and A. A. Romero. 2012. Continuous-flow processes in heterogeneously catalyzed transformations of biomass derivatives into fuels and chemicals. Challenges 3:114–132.
2. Glasnov, T. N., and C. O. Kappe. 2011. The microwave-to-flow paradigm: translating high-temperature batch microwave chemistry to scalable continuous-flow processes. Chem. Eur. J. 17:11956–11968.
3. PBL Netherland Environmental Assessment Agency. Trends in global CO2 emissions. http://edgar.jrc.ec.europa.eu/CO2REPORT2012.pdf (accessed 15 March 2013).
4. Mohan, D., C. U. Pittman, and P. H. Steele. 2006. Pyrolysis of wood/biomass for bio-oil: a critical review. Energy Fuels 20:848–889.
5. Heo, H. S., H. J. Park, Y.-K. Park, C. Ryu, D. J. Suh, Y.-W. Suh, et al. 2010. Bio-oil production from fast pyrolysis of waste furniture sawdust in a fluidized bed. Bioresour. Technol. 101:S91–S96.
6. Cho, M.-H., S.-H. Jung, and J.-S. Kim. 2009. Pyrolysis of mixed plastic wastes for the recovery of benzene, toluene, and xylene (BTX) aromatics in a fluidized bed and chlorine removal by applying various additives. Energy Fuels 24:1389–1395.
7. Kantarelis, E., and A. Zabaniotou. 2009. Valorization of cotton stalks by fast pyrolysis and fixed bed air gasification for syngas production as precursor of second generation biofuels and sustainable agriculture. Bioresour. Technol. 100:942–947.
8. Luque, R., J. A. Menendez, A. Arenillas, and J. Cot. 2012. Microwave-assisted pyrolysis of biomass feedstocks: the way forward? Energy Environ. Sci. 5:5481–5488.
9. Wulff, N., H. Carrer, and S. Pascholati. 2006. Expression and purification of cellulase Xf818 from Xylella fastidiosa in Escherichia coli. Curr. Microbiol. 53:198–203.
10. Atsumi, S., T. Hanai, and J. C. Liao. 2008. Non-fermentative pathways for synthesis of branched-chain higher alcohols as biofuels. Nature 451:86–89.

11. Gustavsson, L., P. Börjesson, B. Johansson, and P. Svenningsson. 1995. Reducing CO2 emissions by substituting biomass for fossil fuels. Energy 20:1097–1113.
12. Lee, S. W., T. Herage, and B. Young. 2004. Emission reduction potential from the combustion of soy methyl ester fuel blended with petroleum distillate fuel. Fuel 83:1607–1613.
13. Gielen, D. J, A. J. M. Bos, M. A. R. C. de Feber, and T. Gerlagh. Biomass for greenhouse gas emission reduction. http://www.ecn.nl/docs/library/report/2000/c00001.pdf (accessed 15 March 2013).
14. Gustavsson, L., J. Holmberg, V. Dornburg, R. Sathre, T. Eggers, K. Mahapatra, et al. 2007. Using biomass for climate change mitigation and oil use reduction. Energy Policy 35:5671–5691.
15. Chen, K., H. Zhang, Y. Miao, M. Jiang, and J. Chen. 2010. Succinic acid production from enzymatic hydrolysate of sake lees using Actinobacillus succinogenes 130Z. Enzyme Microb. Technol. 47:236–240.
16. Oliveira, L. S., and S. F. Adriana. 2009. From solid biowastes to liquid biofuels. Agriculture Issues and Policies Series: 265. Available at: http://www.demec.ufmg.br/disciplinas/eng032-BL/solid_biowastes_liquid_biofuels.pdf (accessed May 2013).
17. Toledano, A., L. Serrano, A. M. Balu, R. Luque, A. Pineda, and J. Labidi. 2013. Fractionation of organosolv lignin from olive tree clippings and its valorization to simple phenolic compounds. ChemSusChem 6:529–536.
18. Du, C., J. Sabirova, W. Soetaert, and C. S. K. Lin. 2012. Polyhydroxyalkanoates production from low-cost sustainable raw materials. Curr. Chem. Biol. 6:14–25.
19. Balu, A. M., V. Budarin, P. S. Shuttleworth, L. A. Pfaltzgraff, K. Waldron, R. Luque, et al. 2012. Valorisation of orange peel residues: waste to biochemicals and nanoporous materials. ChemSusChem 5:1694–1697.
20. Au, E. 2013. Food waste management and practice in Hong Kong in Commercial and Industrial (C&I) Food Waste Recycling Seminar, 8 February 2013, Food Education Association, The Hong Kong Polytechnic University, Hong Kong.
21. Zhang, R., H. M. El-Mashad, K. Hartman, F. Wang, G. Liu, C. Choate, et al. 2006. Characterization of food waste as feedstock for anaerobic digestion. Bioresour. Technol. 98:929–935.
22. Russ, W., and R. Meyer-Pittroff. 2004. Utilizing waste products from the food production and processing industries. Crit. Rev. Food Sci. Nutr. 44:57–62.
23. Kornegay, E. T., G. W. Vander Noot, K. M. Barth, W. S. MacGrath, J. G. Welch, and E. D. Purkhiser. 1965. Nutritive value of garbage as a feed for swine. I. Chemical composition, digestibility and nitrogen utilization of various types of garbage. J. Anim. Sci. 24:319–324.
24. Westendorf, M. L. 1996. Pp. 24–32 in The use of food waste as a feedstuff in swine diets. Proceeding of Food Waste Recycling Symp. Rutgers Coop. Ext., Rutgers Univ.-Cook College, New Brunswick, NJ.
25. Grolleaud, M. 2002. Post-harvest losses: discovering the full story. Overview of the phenomenon of losses during the post-harvest system. FAO, Agro Industries and Post-Harvest Management Service, Rome, Italy.
26. Parfitt, J., M. Barthel, and S. Macnaughton. 2010. Food waste within food supply chains: quantification and potential for change to 2050. Philos. Trans. R. Soc. Lond. B Biol. Sci. 365:3065–3081.

27. Gustavsson, J., C. Cederberg, U. Sonesson, R. van Otterdijk, and A. Meybeck. 2011. Global food losses and food waste: extent, causes and prevention. FAO, Rome, Italy.
28. Tatsi, A., and A. Zouboulis. 2002. A field investigation of the quantity and quality of leachate from a municipal solid waste landfill in a Mediterranean climate (Thessaloniki, Greece). Adv. Environ. Res. 6:207–219.
29. EPD (Environmental Protection Department of HKSAR). Monitoring of solid waste in Hong Kong 2011. https://www.wastereduction.gov.hk/chi/materials/info/msw2011tc.pdf (accessed October 2012).
30. Abu-Rukah, Y., and O. Al-Kofahi. 2001. The assessment of the effect of landfill leachate on ground-water quality—a case study El-Akader landfill site-north Jordan. J. Arid Environ. 49:615–630.
31. Pfaltzgraff, L. A., M. De bruyn, E. C. Cooper, V. Budarin, and J. H. Clark. 2013. Food waste biomass: a resource for high-value chemicals. Green Chem. 15:307–314.
32. Toledano, A., L. Serrano, J. Labidi, A. Pineda, A. M. Balu, and R. Luque. 2013. Heterogeneously catalysed mild hydrogenolytic depolymerisation of lignin under microwave irradiation with hydrogen-donating solvents. ChemCatChem 5:977–985.
33. Toledano, A., L. Serrano, and J. Labidi. 2012. Process for olive tree pruning lignin revalorisation. Chem. Eng. J. 193–194:396–403.
34. Toledano, A., L. Serrano, A. Pineda, A. A. Romero, J. Labidi, and R. Luque. 2013. Microwave-assisted depolymerisation of organosolv lignin via mild hydrogen-free hydrogenolysis: catalyst screening. Appl. Catal. B. doi: 10.1016/j.apcatb.2012.10.015
35. Llorach, R., J. C. Espín, F. A. Tomás-Barberán, and F. Ferreres. 2003. Valorization of cauliflower (Brassica oleracea L. var. botrytis) by-products as a source of antioxidant phenolics. J. Agric. Food Chem. 51:2181–2187.
36. González-Sáiz, J. M., C. Pizarro, I. Esteban-Díez, O. Ramírez, C. J. González-Navarro, M. J. Sáiz-Abajo, et al. 2007. Monitoring of alcoholic fermentation of onion juice by NIR spectroscopy: valorization of worthless onions. J. Agric. Food Chem. 55:2930–2936.
37. Dong, L.-M., X.-P. Yan, Y. Li, Y. Jiang, S.-W. Wang, and D.-Q. Jiang. 2004. On-line coupling of flow injection displacement sorption preconcentration to high-performance liquid chromatography for speciation analysis of mercury in seafood. J. Chromatogr. A 1036:119–125.
38. Cheng, Y., L. Fan, H. Chen, X. Chen, and Z. Hu. 2005. Method for on-line derivatization and separation of aspartic acid enantiomer in pharmaceuticals application by the coupling of flow injection with micellar electrokinetic chromatography. J. Chromatogr. A 1072:259–265.
39. de Boer, A. R., T. Letzel, D. A. van Elswijk, H. Lingeman, W. M. Niessen, and H. Irth. 2004. On-line coupling of high-performance liquid chromatography to a continuous-flow enzyme assay based on electrospray ionization mass spectrometry. Anal. Chem. 76:3155–3161.
40. Stewart, J. J., T. Akiyama, C. Chapple, J. Ralph, and S. D. Mansfield. 2009. The effects on lignin structure of overexpression of ferulate 5-hydroxylase in hybrid poplar. Plant Physiol. 150:621–635.
41. Sahu, R., and P. L. Dhepe. 2012. A one-pot method for the selective conversion of hemicellulose from crop waste into C5 sugars and furfural by using solid acid catalysts. ChemSusChem 5:751–761.

42. Chakraborty, R., S. Bepari, and A. Banerjee. 2010. Transesterification of soybean oil catalyzed by fly ash and egg shell derived solid catalysts. Chem. Eng. J. 165:798–805.
43. Fu, B., L. Gao, L. Niu, R. Wei, and G. Xiao. 2009. Biodiesel from waste cooking oil via heterogeneous superacid catalyst SO42−/ZrO2. Energy Fuels 23:569–572.
44. Clark, J. H., V. Budarin, T. Dugmore, R. Luque, D. J. Macquarrie, and V. Strelko. 2008. Catalytic performance of carbonaceous materials in the esterification of succinic acid. Catal. Commun. 9:1709–1714.
45. Luque, R., A. Pineda, J. C. Colmenares, J. M. Campelo, A. A. Romero, J. C. Serrano-Ruiz, et al. 2012. Carbonaceous residues from biomass gasification as catalysts for biodiesel production. J. Nat. Gas Chem. 21:246–250.
46. Abbot, A. P., R. C. Harris, K. S. Ryder, C. D'Agostino, L. F. Gladden, and M. D. Mantle. 2011. Glycerol eutectics as sustainable solvent systems. Green Chem. 13:82–90.
47. Carriazo, D., M. C. Serrano, M. C. Gutierrez, M. L. Ferrer, and F. del Monte. 2012. Deep eutectic solvents playing multiple roles in the synthesis of polymers and related materials. Chem. Soc. Rev. 41:4996–5014.
48. Zhang, Q., K. De Oliveira Vigier, S. Royer, and F. Jerome. 2012. Deep eutectic solvents: syntheses, properties and applications. Chem. Soc. Rev. 41:7108.
49. Russ, C., and B. König. 2012. Low melting mixtures in organic synthesis- an alternative to ionic liquids? Green Chem. 14:2969–2982.
50. Serrano-Ruiz, J. C., J. M. Campelo, M. Francavilla, C. Menendez, A. B. Garcia, A. A. Romero, et al. 2012. Efficient microwave-assisted production of furfural from C5 sugars in aqueous media catalysed by Brönsted acidic ionic liquids. Catal. Sci. Technol. 2:1828–1832.
51. Zhang, Z., Q. Wang, H. Xie, W. Liu, and Z. K. Zhao. 2011. Catalytic conversion of carbohydrates into 5-hydroxymethylfurfural by germanium (IV) chloride in ionic liquids. ChemSusChem 4:131–138.
52. Colmenares, J. C., R. Luque, J. M. Campelo, F. Colmenares, Z. Karpiński, and A. A. Romero. 2009. Nanostructured photocatalysts and their applications in the photocatalytic transformation of lignocellulosic biomass: an overview. Materials 2:2228–2258.
53. Stillings, R. A., and R. J. V. Nostrand. 1944. The action of ultraviolet light upon cellulose. I. Irradiation effects. II. Post-irradiation effects1. J. Am. Chem. Soc. 66:753–760.
54. Ai, Z., W. Ho, and S. Lee. 2011. Efficient visible light photocatalytic removal of NO with BiOBr-graphene nanocomposites. J. Phys. Chem. 115:25330–25337.
55. Ismail, A. A., and D. W. Bahnemann. 2011. Mesostructured Pt/TiO2 nanocomposites as highly active photocatalysts for the photooxidation of dichloroacetic acid. J. Phys. Chem. 115:5784–5791.
56. Bernardo, E. C. 2008. Solid-waste management practices of households in Manila, Philippines. Ann. NY Acad. Sci. 1140:420–424.
57. Office of the Chief Executive. 2013. Policy Address, 2013 (Office of the Chief Executive). The Hong Kong Government Special Administrative Region (HKSAR), Hong Kong. Available at http://www.policyaddress.gov.hk/2013/eng/p142.html (accessed 16 January 2013).

58. Takata, M., K. Fukushima, N. Kino-Kimata, N. Nagao, C. Niwa, and T. Toda. 2012. The effects of recycling loops in food waste management in Japan: based on the environmental and economic evaluation of food recycling. Sci. Total Environ. 432:309–317.
59. Bernstad, A., and J. la Cour Jansen. 2012. Separate collection of household food waste for anaerobic degradation – Comparison of different techniques from a systems perspective. Waste Manage. (Oxford) 32:806–815.
60. Zhang, B., L.-L. Zhang, S.-C. Zhang, H.-Z. Shi, and W.-M. Cai. 2005. The influence of pH on hydrolysis and acidogenesis of kitchen wastes in two-phase anaerobic digestion. Environ. Technol. 26:329–340.
61. Zhang, A. Y., Z. Sun, C. C. J. Leung, W. Han, K. Y. Lau, M. Li, et al. 2013. Valorisation of bakery waste for succinic acid production. Green Chem. 15:690–695.
62. Van-Thuoc, D., J. Quillaguamán, G. Mamo, and B. Mattiasson. 2008. Utilization of agricultural residues for poly(3-hydroxybutyrate) production by Halomonas boliviensis LC1. J. Appl. Microbiol. 104:420–428.
63. Du, C., S. K. C. Lin, A. Koutinas, R. Wang, P. Dorado, and C. Webb. 2008. A wheat biorefining strategy based on solid-state fermentation for fermentative production of succinic acid. Bioresour. Technol. 99:8310–8315.
64. Dorado, M. P., S. K. C. Lin, A. Koutinas, C. Du, R. Wang, and C. Webb. 2009. Cereal-based biorefinery development: utilisation of wheat milling by-products for the production of succinic acid. J. Biotechnol. 143:51–59.
65. Lin, C. S. K., R. Luque, J. H. Clark, C. Webb, and C. Du. 2012. Wheat-based biorefining strategy for fermentative production and chemical transformations of succinic acid. Biofuels Bioprod. Biorefin. 6:88–104.
66. Leung, C. C. J., A. S. Y. Cheung, A. Y.-Z. Zhang, K. F. Lam, and C. S. K. Lin. 2012. Utilisation of waste bread for fermentative succinic acid production. Biochem. Eng. J. 65:10–15.
67. García, I. L., J. A. López, M. P. Dorado, N. Kopsahelis, M. Alexandri, S. Papanikolaou, et al. 2013. Evaluation of by-products from the biodiesel industry as fermentation feedstock for poly(3-hydroxybutyrate-co-3-hydroxyvalerate) production by Cupriavidus necator. Bioresour. Technol. 130:16–22.
68. Delgado, R., A. J. Castro, and M. Vázquez. 2009. A kinetic assessment of the enzymatic hydrolysis of potato (Solanum tuberosum). LWT Food Sci. Technol. 42:797–804.
69. Yu, J., Z. Li, Q. Ye, Y. Yang, and S. Chen. 2010. Development of succinic acid production from corncob hydrolysate by Actinobacillus succinogenes. J. Ind. Microbiol. Biotechnol. 37:1033–1040.
70. Chen, K., H. Zhang, Y. Miao, P. Wei, and J. Chen. 2011. Simultaneous saccharification and fermentation of acid-pretreated rapeseed meal for succinic acid production using Actinobacillus succinogenes. Enzyme Microb. Technol. 48:339–344.
71. Wang, R., L. C. Godoy, S. M. Shaarani, M. Melikoglu, A. Koutinas, and C. Webb. 2009. Improving wheat flour hydrolysis by an enzyme mixture from solid state fungal fermentation. Enzyme Microb. Technol. 44:223–228.
72. Li, Q., J. Siles, and I. Thompson. 2010. Succinic acid production from orange peel and wheat straw by batch fermentations of Fibrobacter succinogenes S85. Appl. Microbiol. Biotechnol. 88:671–678.

73. Quillaguamán, J., R. Hatti-Kaul, B. Mattiasson, M. T. Alvarez, and O. Delgado. 2004. Halomonas boliviensis sp. nov., an alkalitolerant, moderate halophile isolated from soil around a Bolivian hypersaline lake. Int. J. Syst. Evol. Microbiol. 54:721–725.
74. Van-Thuoc, D., H. Guzmán, J. Quillaguamán, and R. Hatti-Kaul. 2010. High productivity of ectoines by Halomonas boliviensis using a combined two-step fed-batch culture and milking process. J. Biotechnol. 147:46–51.
75. Arancon, R. A., H. R. Barros Jr., A. M. Balu, C. Vargas, and R. Luque. 2011. Valorisation of corncob residues to functionalised porous carbonaceous materials for the simultaneous esterification/transesterification of waste oils. Green Chem. 13:3162–3167.
76. Cabeza, L., M. M. Taylor, G. L. DiMaio, E. Brown, W. N. Marmer, R. Carrió, et al. 1998. Processing of leather waste: pilot scale studies on chrome shavings. Isolation of potentially valuable protein products and chromium. Waste Manage. (Oxford) 18:211–218.
77. Gelse, K., E. Pöschl, and T. Aigner. 2003. Collagens—structure, function, and biosynthesis. Adv. Drug Deliv. Rev. 55:1531–1546.
78. Mata, T. M., A. A. Martins, and N. S. Caetano. 2013. Valorization of waste frying oils and animal fats for biodiesel production. Pp. 671–693 in J. W. Lee, ed. Advanced biofuels and bioproducts. Springer, The Netherlands.
79. Reis, R. L., N. M. Neves, J. F. Mano, M. E. Gomes, A. P. Marques, and H. S. Azevedo. 2008. Natural based polymers for biomedical applications. Woodhead Publishing, CRC Press, Cambridge, U.K.
80. Catalina, M., J. Cot, M. Borras, J. de Lapuente, J. González, A. M. Balu, et al. 2013. From waste to healing biopolymers: biomedical applications of bio-collagenic materials extracted from industrial leather residues in wound healing. Materials 6:1599–1607.
81. Catalina, M., J. Cot, M. Borras, J. de Lapuente, J. González, A. M. Balu, et al. 2013. From waste to healing biopolymers: biomedical applications of bio-collagenic materials extracted from industrial leather residues in wound healing. Materials 6:1599–1607.
82. Langmaier, F., P. Mokrejs, R. Karnas, M. Mládek, and K. Kolomazník. 2006. Modification of chrome-tanned leather waste hydrolysate with epichlorhydrin. J. Soc. Leather Technol. Chem. 90:29–34.
83. Brown, E., C. Thompson, and M. M. Taylor. 1994. Molecular size and conformation of protein recovered from chrome shavings. J. Am. Leather Chem. Assoc. 89:215–220.
84. Yu, J., Y. Hai, and B. Cheng. 2011. Enhanced photocatalytic H2-production activity of TiO2 by Ni(OH)2 cluster modification. J. Phys. Chem. C 115:4953–4958.
85. Liang, Y. T., B. K. Vijayan, K. A. Gray, and M. C. Hersam. 2011. Minimizing graphene defects enhances titania nanocomposite-based photocatalytic reduction of CO2 for improved solar fuel production. Nano Lett. 11:2865–2870.
86. Polshettiwar, V., R. Luque, A. Fihri, H. Zhu, M. Bouhrara, and J. M. Basset. 2011. Magnetically recoverable nanocatalysts. Cheminform 42:3036–3075.
87. Liu, J., S. Z. Qiao, Q. H. Hu, and G. Q. Lu. 2011. Magnetic nanocomposites with mesoporous structures: synthesis and applications. Small 7:425–443.

88. Daskalaki, V. M., M. Antoniadou, G. Li Puma, D. I. Kondarides, and P. Lianos. 2010. Solar light-responsive Pt/CdS/TiO2 photocatalysts for hydrogen production and simultaneous degradation of inorganic or organic sacrificial agents in wastewater. Environ. Sci. Technol. 44:7200–7205.
89. Balu, A. M., B. Baruwati, E. Serrano, J. Cot, J. Garcia-Martinez, R. S. Varma, et al. 2011. Magnetically separable nanocomposites with photocatalytic activity under visible light for the selective transformation of biomass-derived platform molecules. Green Chem. 13:2750–2758.

PART II

TREATMENTS AND PRETREATMENTS

CHAPTER 3

Optimization of Anaerobic Waste Management Systems

ETTORE TRULLI, VINCENZO TORRETTA, ELENA CRISTINA RADA, AND MARCO RAGAZZI

3.1 THE CONSISTENCY OF BIO-WASTE FLOW

The organic waste produced in urban, agricultural and industrial environments mainly comes from:

- municipal solid waste (MSW): organic materials selected from unsorted MSW; organic materials resulting from selective collection (SC);
- wastewater treatment: primary sludge; biological sludge;
- agri-food industry: liquid and solid waste and sludge;
- livestock activities: sewage and sludge.

Selective collection for recycling and the selection of unsorted waste produce different types of organic material flows.

In the management of unsorted waste, mechanical selection obtains a material with a high organic content and good characteristics of putrescence.

Through sieving, two streams of material with different characteristics are obtained. The oversized fraction consists essentially of dry fractions, such as paper and plastic, characterized by a low content of putrescible materials that can be recovered, for example, by incineration. The undersized fraction is the prevailing content of inert materials and organic matter, can be used for the recovery of materials.

In Italy varying amounts of waste are subjected to mechanical biological treatment (MBT), e.g. between 7,628,156 tonnes in 2009 to 9,144,768 tonnes in 2013. With regard to 2013, the amount reported the quantity of 7,194,760 consisted of undifferentiated waste, and the remaining was pretreated waste (782,951) and special waste (I.S.P.R.A., 2014).

The quality of the selected organic fraction is strongly influenced by the composition of the waste. There is a high solid content in organic matter, and the content of volatile solids does not exceed 50% of the total solids. Inert materials are not easily separated if specific steps are not provided that affect subsequent processes such composting or biological stabilization. During co-digestion, the high content of inert materials affects the quality of the digested mixture, making it unsuitable for agronomic recovery.

The SC of the wet organic fraction comes mainly from industrial users and partly from households. The characteristics vary considerably. Table 1 shows the data collection in Italy regarding the organic fraction, i.e. the sum of the "wet" and "green", fractions from 2009 to 2013 (I.S.P.R.A., 2014).

TABLE 1: Organic fraction SC (wet and green) in Italy (2009-2013 (I.S.P.R.A., 2014).

Typology	Unit	2009	2010	2011	2012	2013
Total	t/year	10776.7	11452.6	11848	11992.3	12508.9
Organic fraction	t/year	3743.7	4186.8	4500.8	4813.4	5223.5
	%	34.7	36.6	38.0	40.1	41.8

The material collected has a dry content of between 10% and 25%; the content of volatile solids (VS) varies in the range 85-90% of the total solids (TS). The macro-nutrients are typically 2-3% of the TS. The storage phase can lead to significant changes in the quality of the material following the onset of natural fermentation. The material has characteristics that are usually optimal for the direct application of anaerobic processes.

Additional organic materials include the sludge produced in wastewater treatment plants (WWTPs), whose quality and quantity depend on the characteristics of the wastewater and the treatment adopted.

A significant contribution can be made by farms and companies involved in animal husbandry, food processing, and agricultural products. The waste produced by livestock activities (cattle and pigs, poultry breeding) are for the most part made up of the metabolic waste of animals (Al-Dahhan et al., 2005; Moletta, 2005; Liden and Alvarez, 2008).

3.2 PLANNING ELEMENTS

The anaerobic digestion of organic waste is an established practice. Knowledge gained in the scientific and technological fields has led to improvements in the stability of the treated material which at the same time produces biogas.

Table 2 shows the quantities of waste treated in anaerobic digestion plants in 2011-2013 in Italy.

TABLE 2: Waste treated in anaerobic plants in Italy (2011-2013) (I.S.P.R.A., 2014).

Typology of waste	2011		2012		2013	
	t/year	%	t/year	%	t/year	%
Organic fraction from SC	447.5	60.6	571.9	55.3	526.9	50.5
Sludge	211.3	28.6	257.4	24.9	257.1	24.6
Waste from agro-industry	79.5	10.8	204.4	19.8	259.3	24.9
Total	738.3	100.0	1033.7	100.0	1043.3	100.0

Anaerobic processes have a wide field of application (Figure 1).

The mainly "liquid" matrix phase is characterized by a TS content that does not exceed 10%. The liquid phase has loads in the order of 2÷4 kgVS/m^3 day. "Semi-solid" and "solid" phases treat mixtures of waste with a solid content of between 10 and 40% (Vandevivere et al., 2002).

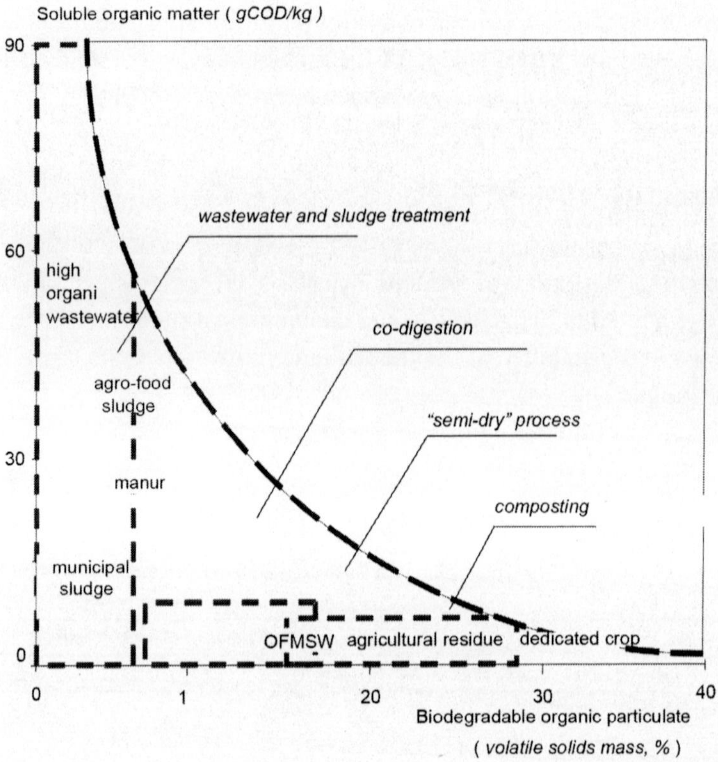

FIGURE 1: Organic waste subjected to anaerobic.

The combined treatment of solid waste and sewage sludge does not lead to significant changes in the biochemical and microbiological process, however it a suitable reactor is required, equipped with lifting and mixing devices. Biologically inert material needs to be separated, followed by the shredding of solids and mixing in order to obtain a homogeneous material. The effectiveness of the process and the environmental impact always need to be taken into consideration.

In relation to the nature of the intervention, the main constraints are:

- location of production and facilities;
- nature and composition of organic matrices;
- adoptable technologies;
- destination of the product;
- regulatory constraints;
- sustainable costs.

The feasibility of the systems is verified by a technical-economic analysis of intervention scenarios. The feasible solutions (depending on the type of plant) are:

- in situ treatment in plants for individual companies;
- treatment plants in a consortium for several companies;
- centralized municipal waste treatment plants;
- centralized treatment at dedicated plants.

Figure 2 illustrates a scenario for the integrated treatment and disposal of bio-waste.

Where feasible, a centralized treatment leads to better yields in terms of biogas production and energy consumption plus a reduction in operating costs and investments, though more complex in terms of management complexity. The increase in energy yields depends on the scale, the possibility of feeding the digester with biomass characterized by a higher potential of biogas, the size of plant smaller at constant load, and the use of larger cogeneration plants. However, transport of the biomass is costly both in financial and environmental terms. For plants already in operation, the organic treatable loads need to be assessed.

Regarding the management of sewage sludge, separating primary from biological sludge is generally a good solution.

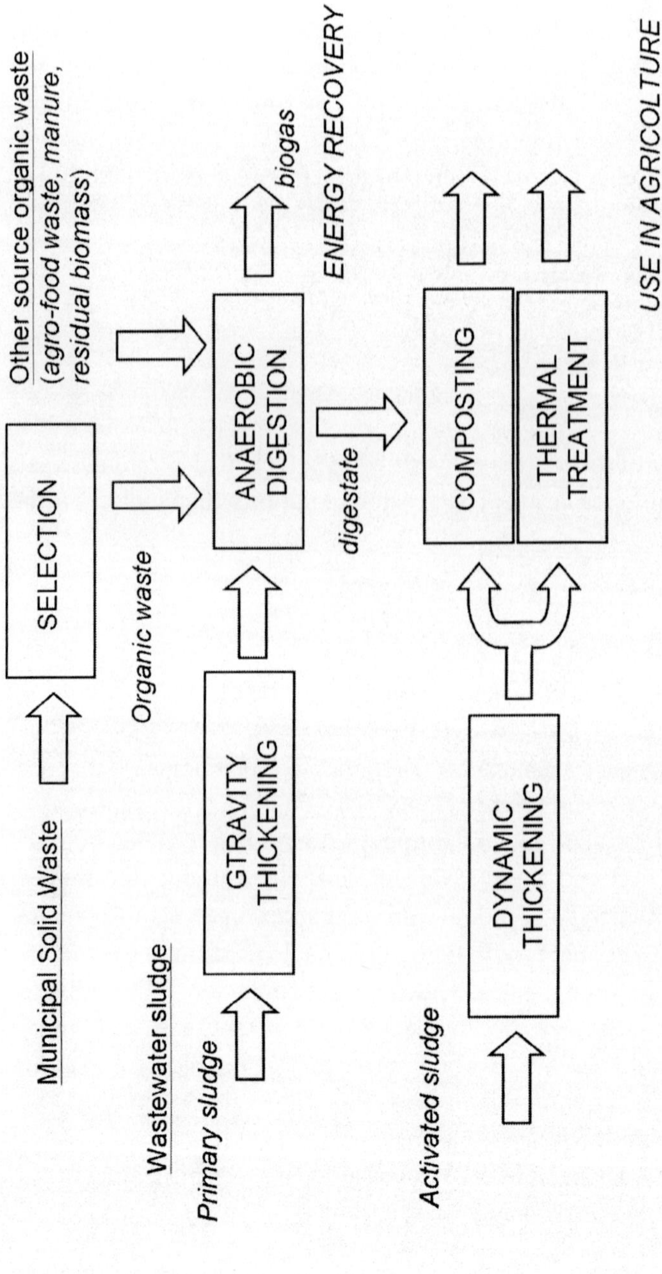

FIGURE 2: Integrated process for treating biodegradable organic waste.

The type of process to use is subject to an analysis of the nature of the mixture (density, settleability, tendency of solids to coalesce or to be adsorbed), the physical form of the organic load (soluble and bound to the organic particulate), and the affinity of the biomass to the substrate.

The system configurations and technological anaerobic applications are classified according to:

- feeding operating modes: continuous or discontinuous;
- conditions for maintenance of biomass: suspended or adherent;
- solid content of the substrate: wet or dry process;
- temperature conditions: mesophilic (35-37 °C) or thermophilic (50-57 °C);
- operation of the reactor: complete mixing, flow piston;
- development of the biological process: just one phase or separate stages.

The digestion of the organic fraction is conducted generally in one or two-stage systems. The plants commonly consist of completely mixed reactors. Completely mixed reactors offer a good performance for the treatment of wastewater with a prevailing content of slowly hydrolyzable insoluble solids.

The reactors with a "piston" flow are used in the treatment of mixtures characterized by a high density. In this case verification of the handling and blending conditions of the mixture in digestion is essential. For the treatment of predominantly dissolved organic loads, mud bed reactors work well, while in the absence of suspended solids, fluidized and expanded bed reactors are highly efficient.

The anaerobic treatment of organic substrates should be carefully evaluated. It is suitable in areas where the plants interact with production centers or farms with a sufficient area available for production in energy crops. The mixing of "pure" substrates with sludge and waste with "undesirable substances" such as inorganic and microbiological contaminants, needs to be verified.

Integrated anaerobic-aerobic methods are now widely applied. Although an evaluation of the investment and operation costs is essential, this solution is considered highly valuable both for the construction of new plants and the up-grading of existing ones. These processes are suitable to treat organic materials with a high degree of fermentation. In such systems, composting leads to the refinement of the digestate.

The process is conducted in reactors, drum or bio-cells, with mechanized processes, or in covered areas in especially equipped sheds. The main purpose of these systems is to better control the operating conditions, which are not influenced by meteorological conditions, and to limit the impact of gaseous emissions. Based on the mixing and handling of the material to be treated, the systems are divided into static and dynamic.

3.3 OPTIMIZATION OF THE PERFORMANCE OF ANAEROBIC SYSTEMS

The increasing constraints holding back the disposal of organic waste require the research and implementation of the best available technologies. The optimization of anaerobic plants can be achieved by:

- speeding up biological activities;
- reducing the effluent solid mass;
- improving the quality of solids;
- reducing energy consumption;
- increasing the treatment capacity of the digesters;
- implementing management functions;
- getting a faster return on investment.

3.3.1 MANAGEMENT INTERVENTIONS

In the operation of anaerobic plants, the following problems need managing: the control of odors, the formation of foam, materials and fat floatation, stratification and sedimentation, instability of the process with reduced pH, poor degradation of organic substances and volatile solids, poor quality, and low amount of biogas.

Operating conditions can be optimized through uniform loading, suitable mixing conditions, and adequate heating. Feeding solids ahead of the heat exchanger or recirculation pump and automatically removing foam and floating substances can be arranged (Rada et al., 2013).

Particular attention must be paid to the preparation of the mixtures to be treated (Ragazzi et al.,2014; Trulli and Torretta, 2015). In addition, the input materials may contain impurities such as sand, glass, metals, plastics, wood residues, soil, straw, bristles, and inhibitory substances, such as ammonia, antibiotics, disinfectants and pesticides.

These impurities are difficult to control once they have been introduced into the digester and often cause failures in the process (Demirer and Chen, 2005; Capela et al., 2008; Lindorfer et al., 2008). These materials must be removed in advance by screening, grit removal, thickening high-blend liquids, and grinding solid and fibrous components.

The high content of water in the substrates results in an increase in costs due to the transfer of heat. The mixtures with a high solid content modify the fluid dynamics of the process or hinder the mixing and material handling equipment.

During the instability phases, controlling the operating parameters (in the digester and in the produced biogas: temperature, pH, alkalinity, biogas quality, volatile acids) is important as well as checking the hydraulic and solid residence times and the applied organic loads.

Note that the presence of toxic compounds in the substrate and inhibiting anaerobic biomass, must be carefully evaluated, also taking into account the capacity to acclimatize the microorganisms (Hartman and Ahring, 2005). In the thermophilic digestion of manure, the acclimatization of biomass has helped to alleviate the effects of inhibition due to high concentrations of ammonia. In the co-digestion of municipal waste, the limit value of the ratio COD/N was estimated as about 50.

3.3.2 INTERVENTIONS ON PLANT SCHEMES

A greater reduction in solids reduces the mass to be treated in the post-treatment and to be transported to the final disposal, as well as decreasing the consumption of polymers and environmental impacts associated with the production and transport of chemical reagents.

In order to improve the performance of anaerobic digestion, it is possible to increase the hydrolysis by thermal and chemical processes (Ward et al., 2008). Further improvements can also be obtained through the optimization of processes and enzymatic catalysis. Pathogens can be removed by thermal drying, thermophilic aerobic digestion, and pasteurization.

3.3.2.1 INCREASING SOLID RESIDENCE TIME

The solid residence time in conventional complete mixing installations can be made independent from the hydraulic residence time, by separating the solids. Several technologies are available. Conventional "in contact" processes use sedimentation and thickening. Surveys recently conducted in full-scale plants also tested floatation processes, with anoxic gas or dissolved air, gravity presses, and filtration membranes.

From an operational point of view, it is possible to operate with higher concentrations of solids, which are useful to dab overload conditions. The main disadvantages are the complexity of the operation and the investment required, which are higher for larger plant surfaces and for the solid separation unit.

The kinetics of microorganisms recirculated into the digester require further technical and scientific evaluations, in particular the strictly anaerobic species, which are more sensitive to environmental conditions, exposed to aerobic conditions and lower temperatures. Some studies show that methanogenic bacteria can survive temporary changes in environmental conditions.

3.3.2.2 USING THERMOPHILIC CONDITIONS

Temperatures above 50° enable operations with faster kinetics, by increasing the destruction of VS, and reducing the content of pathogens. However operating temperatures of above 57°C may lead to thermal shock (Ndegwa et al. 2008; Kim et al., 2008). A two-stage thermophilic-mesophilic process has the same advantages as thermophilic digestion, and does not involve the diffusion of odors resulting from the increased conversion of volatile acids. However, thermophilic process has the following disadvantages:

- higher energy consumption for heating the mixture in digestion and keeping the digester at operating temperatures;
- higher ammonia concentration in the effluent liquid and the need for nitrogen control of the treatment plants;
- more intense smell;
- increase in moisture in the gas and increase in the condensate in the digester;

- mechanical problems caused by the inlay of heat exchangers;
- thermal stress on the digester construction material.

3.3.2.3 PROCESS PHASES

The use of two reactors in series, which provide optimal environments for the acidifying methanogenic phases, can lead to a greater destruction of VS: the first acidifying digester operates with a solid residence time lower than 1-2 days, and is subject to a high organic load; the second methanogenic digester operates with a longer residence time, of about 10 days. In the first digester, hydrolysis of the solid particulate takes place, and there is a minimum production of biogas, not more than 6-10% of the total, with a methane content of less than 40%. The process leads to a better control of foam and an increase in gas production. The operating parameters need to be controlled and the gases produced in the two reactors need to be managed separately with a different quality (hydrogen sulphide, methane) and different calorific value.

3.4 USE OF ANAEROBIC DIGESTION FOR ENERGY PRODUCTION

The forms of energy recovery consist in the use of biogas as a fuel and the recovery of energy and heat from the incineration of refuse-derived fuel. CHP plants can fit easily both in consortium organizations and centralized systems.

Interventions aimed at the production of energy from renewable sources are currently being implemented by the scientific community. The use of biomass for the production of biogas through anaerobic biological processes has received great attention (Murphy and McKeogh, 2004; Berglund and Borjesson, 2006; Lehtomäki, 2006; Amon et al., 2007; Martinez et al., 2014; Torretta and Trulli, 2015). Economic incentives are provided by national authorities through green certificates, which can be traded on the energy market, and are issued and controlled by the network operator.

An optimal solution for biogas is provided by co-generation systems that combine the production of heat and electricity. This technique uses internal combustion engines and gas microturbines. The internal combustion engines operate using Otto or Diesel cycles. Of the microturbines, turbomachinery the use of a Joule-Brayton cycle, characterized by smaller dimensions, is common. Safety devices are crucial along with the prevention of fire risks and explosions in the treatment phases, accumulation and use of biogas.

In order to use methane from biogas in urban vehicles, it is necessary to purify the biogas with a methane content of about 95%. Combined systems can recover carbon dioxide for technical uses, though they need a well-organized infrastructure network.

The most recent trends related to the production of renewable energy systems involve the use of hydrogen. The hydrogen can be produced by combining organic fermentation with photosynthetic bacteria, which use the acids produced in the anaerobic phase.

One of the major problems is the separation of hydrogen from the biogas with membrane separation. Fuel cells should be able to provide a viable solution to this problem in the near future.

3.5 USE OF DIGESTATE

The options include the reuse of treated organic waste on agricultural land, through direct contribution or as a constituent of quality compost.

The digested materials may be suitable for specific uses in agriculture and forestry, such as soil conditioners by providing nutrients and organic matter and improving soil structure and water retention. The treated mixtures can be used as they are or after aerobic composting. Once the fibrous material has been separated from the digested solid fraction, it can be used in the redevelopment of specific sites. For mixtures produced starting from MSW, the presence of unwanted materials can be a great problem. In these systems, the energy of the biogas can meet the energy demand to dehydrate the digestate.

Knowledge of integrated systems needs to be increased, for example to areas related to the prevention of infectious processes. The liquid phase obtained from the separation of the digested mixture, which is characterized by a high content of nitrogen, can be used as a fertilizer. It has better characteristics compared to sludge in terms of handling and distribution on crops, and by virtue of the minimization of the losses due to volatilization. In order to improve the composition, it can be integrated with mineral and organic additives.

The use of a soil improver with a high nitrogen content may be limited by legislation on the control of nitrates in vulnerable zones and by good agricultural practice. The current situation of the market for digested material needs to be improved by considering the trend in the cost of fertilizers, the evolution of the practices for the disposal of sewage sludge, and EU legislation.

3.6 CONTROLLING THE NITROGEN IN THE EFFLUENTS

Controlling the nutrient content and of nitrogen in the supernatant of the anaerobic digestion is key given the current public focus on protecting water resources from excessive inputs of nutrients, as well as ensuring the quality of air, and reducing global warming by cutting emissions of ammonia and nitrogen oxides.

The main current technologies for nitrogen removal are:

- physical and thermal processes: solid-liquid separation; ammonia stripping; evaporation / concentration; membrane filtration;
- chemical processes: chemical precipitation of ammonium salts (struvite); adsorption on zeolites;
- biological processes: nitrification-denitrification by conventional and innovative process; phytoremediation.

3.6.1 PROCESSES BASED ON CHEMICAL PRECIPITATION

Nitrogen removal processes for recovering materials for use as fertilizers are currently of considerable interest and particularly those based on the crystallization of struvite, magnesium ammonium phosphate (MAP) (Uludag-Demirer et al., 2005).

In reactor suitable conditions are operated to realize struvite precipitation. Struvite precipitates at pH values greater than 8. The process develops through various dosages: the bases (calcium hydroxide, sodium or magnesium), carbon dioxide, chloride and magnesium hydroxide. In the presence of appropriate concentrations of magnesium and phosphorus, nitrogen conversion reaches a yield of about 90%, with a stoichiometric ratio of the molar fractions of 1:1:1. A slight excess of magnesium and phosphorus is required. However, further increases do not give higher yields. Temperatures between 25 and 40 °C do not significantly affect the removal of the ammonia at values greater than 90%. The influence of pH is important: the highest yields have been observed with values in the range 8.5-9.0. The rate of reaction is very high and the process is completed in a few minutes, allowing the adoption of relatively low volumes of the reactor. Struvite crystals from the impurities of the matrices can be washed by dissolving the precipitate in acid solution, removing by centrifugation, and precipitating the clarified supernatant by pH adjustment with caustic soda. Experimental investigations conducted with two-stage processes have led to a recovery of the material of over 85% (Celen and Türker, 2001).

3.6.2 INNOVATIVE BIOLOGICAL PROCESSES

Advanced engineering solutions based on biological processes that realize the oxidation and reduction of nitrogen in concentrated flows are increasingly being sought after. The partial-flow nitrification and autotrophic denitrification play an important role by utilizing the activity of specialized bacterial populations. The microbiological aspects are still subject to experimentation but the process allows an ammonia nitrogen removal of up to 90% using limited quantities of oxygen in the nitrification and organic carbon in the denitrification compared to conventional processes.

Partial nitrification is used for treating wastewater with high concentrations of nitrogen. The SHARON® process (Single reactor High-activity Ammonium Removal Over Nitrite) is currently considered to be the best solution (Hellinga et al., 1998, Van Hulle et al., 2007).

At relatively high temperatures (25÷30 °C) the nitrite-oxidizing biomass reduced growth rates compared to ammonium-oxidizing. In a completely mixed reactor without recirculation of sludge, with a short residence time, usually 1÷2 days, the nitrite-oxidizing biomass can be washed out. Operating an intermittent aeration, denitrification and pH control were obtained.

A sequencing batch reactor (SBR) represents a suitable plant system. The operating conditions must allow a growth rate of the ammonium-oxidizing bacteria that is sufficiently high to maintain a high biomass concentration.

Autotrophic denitrification is promoted by microorganisms growing in the absence of oxygen with a very low growth rate. The ammonia nitrogen is transformed into nitrogen gas using nitrate as electron acceptors. Among the most interesting plant applications, ANAMMOX® (ANaerobic AMMonium OXidation) consumes less oxygen and biodegradable organic matter, produces less sludge, and has a high conversion rate of nitrogen than conventional processes (Fux and Siegrist, 2004). The main disadvantages are the low growth rate of bacterial species promoting the process and the potential presence of nitrites in the effluent in concentrations above the regulatory limits.

The SHARON® and ANAMMOX® processes are also used in combined systems: the SHARON® process carries out the first stage by converting a large part of the load ammonia to nitrite; the ANAMMOX® process operates the second stage in which the ammonium and nitrite are converted into nitrogen gas.

3.7 CONCLUSIONS

The optimal management of treatment and disposal of the organic waste requires a modernization of plants and process lay-out as well as improved operations. The technological advances that can be promoted by the applied research and best practice play an important role in improving such operations. Recent legislative guidelines and the increased costs of waste disposal favor the use the anaerobic technologies, as demonstrated by the growing number of plants. Anaerobic digestion operated in an integrated lay-out, combined with auxiliary processes, constitutes a suitable treatment. Although the process is quite stable, there are some limitations in the design and management of the reactors such as the composition of the mixture of organic material to be treated, the toxic effects of substances that may be contained in digestate, and the growth of specialized biomass.

REFERENCES

1. Al-Dahhan M. H., K. Khurscheed, R. Hoffmann, T. Klasson, S.R. Drescher, De Paoli D. (2005). Anaerobic digestion of animal waste: effect of mixing. Bioresource Technology, 96, 1607-1612.
2. Amon B., Amon T., Kryvoruchko V., Zollitsch K. M., Gruber. L. (2007). Biogas production from maize and dairy cattle manure: influence of biomass composition on the methane yield. Agriculture, Ecosystem and Environment. 118, 173-182.
3. Berglund M., Borjesson P. (2006). Assessment of energy performance in the life-cycle of biogas production. Biomass and Bioenergy. 30, 254-266,
4. Capela I., Rodrigues A., Silva F., Nadais H., Arroja L. (2008). Impact of industrial sludge and cattle manure on anaerobic digestion of the OFMSW under mesophilic conditions. Biomass and Bioenergy. 32, 245-251.
5. Çelen I., Türker M. (2001). Recovery of Ammonia as Struvite from Anaerobic Digester Effluents. Environmental Technology. 22, 1263-1272.
6. Demirer G. N., Chen S. (2005). Two-phase anaerobic digestion of unscreened dairy manure. Process Biochemistry. 40, 3542-3549.
7. Fux, C., Siegrist, H. (2004). Nitrogen removal from sludge digester liquids by nitrification/denitrification or partial nitritation/anammox: environmental and economical considerations. Water Science and Technology. 50(10), 19-26.
8. Kim I. S., Chae K. J., Am Jang, Yim S. K. (2008). The effect of digestion temperature and temperature shock on the biogas yield from the mesophilic anaerobic digestion of swine manure. Bioresource Technology. 99, 1-6.
9. Hartman H., Ahring B. (2005). Anaerobic digestion of organic fraction of municipal solid waste: influence of co-digestion with manure. Water Research, 39(8), 1543-1552.
10. Hellinga C., Schellen A., Mulder J., Van Loosdrecht M., J. Heijnen (1998). The SHARON process: an innovative method for nitrogen removal from ammonium-rich wastewater. Water Science and Technology. 37(9), 135-142.
11. I.S.P.R.A. - Istituto Superiore per la Protezione e la Ricerca Ambientale, (2014). Municipal Waste - Report n. 202/2014, Rome, Italy.
12. Lehtomaki A. (2006). Co-digestion of energy crops and crop residues with cow manure. In Biogas production from energy crops and crop residues, Academic dissertation, University of Jyvaskyla, Finland.
13. Liden G., Alvarez R. (2008). Semi-continuous co-digestion of solid slaughterhouse waste, manure and fruit and vegetable waste. Renewable Energy. 33, 726-734.
14. Lindorfer H., Corcoba A., Vasilieva V., Braun R., Kichmayer R. (2008). Doubling the organic loading rate in the co-digestion of energy crops and manure: a full scale case study. Bioresource and Technology. 99(5), 1148-1156.
15. Martinez S., Torretta V., Minguela J., Siñeriz F., Raboni M., Copelli S., Rada E.C., Ragazzi M. (2014). Treatment of slaughterhouse wastewaters using anaerobic filters. Environmental Technology. 35(3), 322-332.
16. Moletta R. (2005). Winery and distillery wastewater treatment by anaerobic digestion. Water Science and Technology. 51(1), 137-144

17. Murphy J.D., McKeogh E. (2004). Technical, economical and environmental analysis of energy production from municipal solid waste. Renewable Energy, 29(7), 1043-1057.
18. Ndegwa P. M., Hamilton D. W., Lalman J. A., Cumba H. J. (2008). Effects of cycle-frequency and temperature on the performance of ASBRs treating swine waste. Bioresource and Technology. 99, 1972-1980.
19. Rada E.C., Ragazzi M., Torretta V. (2013). Laboratory-scale anaerobic sequencing batch reactor for treatment of stillage from fruit distillation. Water Science & Technology. 67(5), 1068-1074.
20. Ragazzi M., Rada E.C., Schiavon M., Torretta V. (2014). Unconventional parameter for a correct biostabilization plant design in agriculture areas. Mitteilungen Klosterneuburg. 64(6), 2-14.
21. Trulli E., V. Torretta (2015). Influence of feeding mixture composition in batch anaerobic co-digestion of stabilized municipal sludge and waste from dairy farms. Environmental Technology. 36(12), 1519-1528.
22. Torretta V., Trulli E. (2015). Energy recovery from co-digestion of complex substrate mixtures: An overview, and case study of biogas production modeling. Biotechnology: An Indian Journal. 11(4), 150-160.
23. Uludag-Demirer S., Demirer G. N., Chen S. (2005). Ammonia removal from anaerobically digested dairy manure by struvite precipitation. Process Biochemistry. 40, 3667-3674.
24. Vandevivere P., De Baere L., Vestraete W. (2002). Types of anaerobic digesters for solid wastes. In "Biomethanization of the organic fraction of municipal solid wastes"; J. Mata-Alvarez Ed., IWA Publ., London, UK.
25. Van Hulle S.W.H., E.I.P. Volcke, J. López Teruel, B. Donckels, M.C.M. van Loosdrecht, P.A. Vanrolleghem (2007). Influence of temperature and pH on the kinetics of the SHARON nitritation process. Journal of Chemical Technology & Biotechnology. 82, 471-480.
26. Ward A., P.Hobbs, P.Holliman, D.Jones (2008). Optimisation of the anaerobic digestion of agricultural resources. Bioresource and Technology, 99, 7928-7940.

CHAPTER 4

Effect of Ultrasonic Pretreatment on Biomethane Potential of Two-Phase Olive Mill Solid Waste: Kinetic Approach and Process Performance

B. RINCÓN, L. BUJALANCE, F. G. FERMOSO, A. MARTÍN, AND R. BORJA

4.1 INTRODUCTION

The two-phase olive mill solid waste (OMSW) is the main waste produced after primary centrifugation in the two-phase olive oil mills. In the two-phase olive oil manufacturing process a horizontally mounted centrifuge is used for primary separation of the olive oil fraction from the vegetable solid material and vegetation water. The resultant olive oil is further washed to remove residual impurities before finally being separated from this wash water in a vertical centrifuge. Therefore, the two-phase olive mills produce three waste streams: wash waters from the initial cleaning of the fruit, an aqueous solid residue called OMSW, and the wash waters generated during the purification of the virgin olive oil [1]. Two-phase OMSW is the main waste produced and has a high organic matter concentration. It is also a very wet waste (60–70% humidity), containing 3% of olive oil and a complex structure formed mainly by lignin (42.6%), cellulose (19.4%), and hemicellulose (35.1%) [2]. These characteristics result in an elevated polluting load. In addition, the quantities of OMSW generated are very large; every year from two to four million tonnes are produced in countries like Spain. Both composition and quantity produced make two-phase OMSW an important environmental problem [3].

Anaerobic digestion of solid wastes is an attractive and established option for solid wastes treatment due to the excellent waste stabilization and high energy recovery [4–7]. The feasibility of the anaerobic digestion of the two-phase OMSW has been already shown [3, 5, 6]. Methane yield coefficients up to 0.244 $LCH_4/g\ COD_{removed}$ were reported [5].

Pretreatments to break complex structures could be a good option to increase the methane yields obtained through anaerobic digestion. Among the most studied pretreatments to improve the hydrolysis and solubilization of complex substrates prior to their anaerobic digestion stands out the use of ultrasounds [8–14]. Ultrasonic pretreatment consists of the application of cyclic sound pressure with a variable frequency to some wastes to disintegrate rigid structures and complex compounds [9, 10]. The chemistry of sonication as a pretreatment tool is quite complex and consists of a combination of shearing, chemical reactions with radicals, pyrolysis, and combustion [13]. During sonication, microbubbles are formed because of high-pressure applications to liquid, which cause violent collapses and high amounts of energy to be released into a small area. Consequently, because of extreme local conditions certain radicals (•HO, •H) can be formed [15, 16]. The radical reactions can degrade volatile compounds by pyrolysis processes taking place in microbubbles [16].

This technology or pretreatment is widely used in industrial plants for WAS in the UK, USA, and Australia achieving a reduction in the volatile solids (VS) content between 30% and 50% and an increase in the biogas production between 40% and 50% [17]. Ultrasound pretreatment has been widely studied for WAS with interesting results and also for other substrates: sewage sludge [9], pulp mill wastewaters [10], hog manure [11], sludge from the pulp, and paper industry [12]. The main target of the ultrasound pretreatment is to disrupt flocks and break the cellular walls making easier the access to the intracellular material for its subsequent degradation. One of the main advantages of the ultrasound pretreatment is that the use of external chemical agents is prevented and, therefore, an increase in the effluent volume is avoided [13].

The effect of the ultrasonication pretreatment for different substrates treated subsequently by anaerobic digestion has been studied during the last years due to an increase in the biogas production and a reduction in the hydraulic retention times needed [8]. Mechanisms of ultrasonic treatment are influenced by four main factors: specific energy, ultrasonic frequency, application time, and the characteristics of the substrate. The increased percentage in biogas production as well as the methane content in the biogas of a sonicated sludge usually increases with the sonication time applied [15, 16].

The increase in specific methane yield is mainly due to the increase in the net surface area of the particles and solubilization of complex organic matter [13]. The increment of the sonication time can reduce the particle size of a substrate [18], but, for very high times of exposition to the ultrasound, the effect of particle size reduction might be stopped and the opposite effect is produced [19]. Initially the flocks are broken, but at high exposure times the intracellular polymeric compounds released might favour a reflocculation process [19, 20]. This might result in negative effect for long exposure times.

The objective of this study was to evaluate the COD solubilization owing to the ultrasonication pretreatment of the two-phase OMSW at different specific energies and application times and to study the influence of this pretreatment on the methane production through biochemical methane potential (BMP) tests. A kinetic study of the different stages in the methane production was also carried out. There are no previous studies in the literature about ultrasound pretreatment of this substrate before its anaerobic digestion process.

4.2 MATERIALS AND METHODS

4.2.1 TWO-PHASE OMSW

The two-phase OMSW used for the experiments was collected from the Experimental Olive Oil Factory located in the "Instituto de la Grasa (CSIC)" of Sevilla, Spain. OMSW was sieved through a 2 mm mesh to remove olive stone pieces; all results are presented for sieved OMSW. The olive variety used was "Lechín" from Sevilla. The main characteristics and composition of the two-phase OMSW are presented in Table 1.

4.2.2 ULTRASOUND PRETREATMENT

Ultrasound pretreatment of two-phase OMSW was performed using ultrasonication equipment Hielscher UP200S (sonotrode Micro tip 7). A maximum power of 200 W (100% amplitude), constant working frequency of 24 kHz, and a constant ultrasound intensity of 5.3 W/cm^2 [21] were used. The ultrasound tip was used in open 100 mL Pyrex glass beakers. Ultrasound pretreatment times of 20, 40, 60, 90, 120, and 180 minutes were studied corresponding to six different specific energies and ultrasound densities [15] (Table 2). All the ultrasound pretreatment experiments were carried out in duplicate and the final results expressed as means.

Two-phase OMSW at 80% (80 g two-phase OMSW: 20 g water) was used for all the experiments with ultrasound pretreatment and without pretreatment. Temperature was not controlled during the ultrasound pretreatment. After ultrasound pretreatment, the samples were cooled to ambient temperature.

4.2.3 BIOCHEMICAL METHANE POTENTIAL (BMP) TESTS

To compare methane yields after the pretreatment, BMP tests were used. BMP tests were carried out in reactors with an effective volume of 250 mL.

Reactors were continuously stirred at 500 rpm and placed in a thermostatic water bath at mesophilic temperature (35 + 2°C).

TABLE 1: Main characteristics and composition of the two-phase OMSW used in the experiments.

Parameters	Values
TS (g/kg)	265 ± 3
VS (g/kg)	228 ± 2
CODt (g O_2/kg)	331 ± 1
CODs (g O_2/kg)	143 ± 3
Ph	4.9 ± 0.2
TA (g $CaCO_3$/kg)	2.5 ± 0.1
AN (g ammoniacal N/kg)	0.3 ± 0.0
TKN (g Kjeldahl N/Kg)	3.6 ± 0.1
Hemicellulose (%)	11.3 ± 0.2
Cellulose (%)	5.2 ± 0.1
Lignin (%)	19.7 ± 0.4
Fat (%)	3.8 ± 0.3

TS: total solids; VS: volatile solids; CODt: total chemical oxygen demand; CODs: soluble chemical oxygen demand; TKN: total Kjeldahl nitrogen; AN: ammoniacal nitrogen; TA: total alkalinity.

The reactors were sealed and the headspace of each flask was filled with nitrogen at the beginning of each assay. The methane produced was measured by liquid displacement passing the biogas through a 3N NaOH solution to capture CO_2 assuming that the remaining gas was methane. The anaerobic digestion experiments were run for a period of 20 days until the accumulated gas production remained essentially unchanged; that is, on the last day production was lower than 2% of the accumulated methane produced. Each experiment was carried out in duplicate.

The inoculum used in the BMP assays was obtained from an industrial anaerobic reactor treating brewery wastewater and operating at mesophilic temperature. The characteristics of the anaerobic inoculum used were pH: 7.5 and VS: 22 g/L.

TABLE 2: Experimental conditions of specific energies and ultrasound densities applied to the two-phase OMSW used in the experiments (80% w/w).

Time (min)	Wet sample (g)	Specific energy (kJ/kgTS)	Ultrasound density (W/kg wet sample)
20	99.6	11367	2008.2
40	107.2	21121	1865.8
60	99.7	34072	2006.6
90	99.3	51284	2013.5
120	99.1	68557	2018.8
180	96.1	106003	2080.9

The inoculum to substrate ratio used was 2 on VS basis. For each flask containing 239 mL of inoculum (with a final concentration of 21 g VS/L), the amount of untreated OMSW or ultrasound pretreated OMSW needed to give the required inoculum : substrate ratio was then added to each test digester. A volume of 0.239 mL of trace element solution was also added to each digester.

The composition of the trace elements solution was $FeCl_2 \cdot 4H_2O$, 2000 mg/L; $CoCl_2 \cdot 6H_2O$, 2000 mg/L; $MnCl_2 \cdot 4H_2O$, 500 mg/L; $AlCl_3 \cdot 6H_2O$, 90 mg/L; $(NH_4)_6Mo_7O_{24} \cdot 4H_2O$, 50 mg/L; H_3BO_3, 50 mg/L; $ZnCl_2$, 50 mg/L; $CuCl_2 \cdot 2H_2O$, 38 mg/L; $NiCl_2 \cdot 6H_2O$, 50 mg/L; $Na_2SeO_3 \cdot 5H_2O$, 194 mg/L; and EDTA, 1000 mg/L. Two reactors with anaerobic inoculum and trace elements solution but without substrate addition were used as controls.

4.2.4 ANALYTICAL METHODS

TS and VS were determined, according to the standard methods 2540B and 2540E, respectively [22]. Total chemical oxygen demand (CODt) was determined as described by Rincón et al. [4], while soluble chemical oxygen demand (CODs) was determined using the closed digestion and the colorimetric standard method 5220D [22]. pH was analysed using a pH-meter model Crison 20 Basic. Total alkalinity (TA) was determined by pH titration to 4.3 [22]. Hemicellulose, cellulose, and lignin were determined according to Van Soest et al. method [23]. Total Kjeldahl nitrogen (TKN) was analysed using a method based on the 4500-N_{org} B of standard methods [22]. Ammoniacal nitrogen was determined by distillation and titration according to the standard method 4500-NH_3 E [22]. Fat was analyzed by the official method of the EEC number 2568/91 (European Community Official Diary, L248/1 of 05.09.1991). All the analyses were carried out in triplicate.

4.3 RESULTS AND DISCUSSION

4.3.1 INFLUENCE OF ULTRASOUND PRETREATMENT ON THE CHARACTERISTICS OF TWO-PHASE OMSW

Table 3 shows the characteristics of the two-phase OMSW after the different ultrasound pretreatments in terms of humidity, TS, VS, CODt, CODs, and COD solubilization. The degree of COD solubilization was calculated from the data of CODs measured after each pretreatment condition tested and CODt initial of the OMSW using the following equation [24, 25]:

$$\text{COD solubilization (\%)} = (\text{CODs}/\text{CODt}) * 100 \quad (1)$$

Although COD solubilization did not show a big variation for the chosen US exposition times (Table 3), the best solubilization levels were achieved for the pretreatments at 90 and 120 min with 57% of solubilization, followed by the treatment at 20 min and 60 min with 56% and 55% of COD solubilization, respectively (Table 3). From the low times assayed (20, 40, and 60 min), 20 min was the chosen time for the BMP tests, as for 40 and 60 min similar solubilizations were virtually achieved. For the same reason from the highest times studied (90, 120, and 180 min) the pretreatment at 90 min was selected. The pretreatment at 180 min was also assayed through BMP to compare the effect of a high exposure to ultrasound pretreatment. Therefore, the BMP tests were assayed for low, that is, 20 min, medium, that is, 90 min, and high, that is, 180 min, exposition times.

TABLE 3: Characteristics of the two-phase OMSW used (80% w/w) after different ultrasound pretreatment times and without pretreatment.

Time (min)	Moisture (%)	TS (g/kg)	VS (g/kg)	CODt (gO_2/kg)	CODs (gO_2/kg)	Solubilization (%)
Untreated OMSW	78.8 ± 0.2	212.0 ± 2.6	182.7 ± 2.3	265.4 ± 0.7	114.7 ± 3.2	43
20	79.9 ± 0.2	206.4 ± 0.2	177.7 ± 1.9	331.3 ± 1.0	148.3 ± 0.2	**56**
40	80.8 ± 0.1	191.8 ± 0.9	164.1 ± 0.2	331.4 ± 0.0	130.9 ± 0.0	49
60	78.6 ± 0.4	213.8 ± 3.6	181.4 ± 5	376.6 ± 0.4	146.8 ± 0.0	55
90	78.3 ± 0.3	217.3 ± 2.9	183.8 ± 0.0	370.0 ± 0.1	151.0 ± 0.0	**57**
120	79.3 ± 0.3	206.8 ± 3.2	173.3 ± 0.0	385.2 ± 0.6	150.6 ± 0.1	57
180	77.9 ± 0.5	221.4 ± 5	188.6 ± 7.4	377.5 ± 0.0	126.1 ± 0.0	**48**

TS: total solids; VS: volatile solids; CODt: total chemical oxygen demand; CODs: soluble chemical oxygen demand.

TABLE 4: Hemicellulose, cellulose, and lignin contents for the untreated two-phase OMSW and ultrasound pretreated OMSW at 20, 90, and 180 min.

Times (min)	Hemicellulose (%)	Cellulose (%)	Lignin (%)
Untreated OMSW	9.0 ± 0.2	4.2 ± 0.1	15.8 ± 0.4
20	10.9 ± 0.3	10.5 ± 0.5	14.4 ± 0.6
90	11.8 ± 0.0	11.6 ± 0.6	14.6 ± 0.1
180	11.6 ± 2.6	11.0 ± 1.2	14.5 ± 1.3

Wang et al. [26] found for WAS that the concentration of soluble COD increased with the increase in the sonication time owing to the breakage to the flocks and the disrupting of cell walls in bacteria that released the extracellular organic compounds. Shimizu et al. [8] also evaluated the solubilization of WAS at different sonication times; they found that a minimum of 30–40 min of ultrasonication time was necessary to achieve 50% of solubilization. The efficiency of ultrasonication as a pretreatment method for hog manure and WAS prior to their anaerobic digestion has been recently evaluated at specific energies of 250–30,000 kJ/kg TS [11]. The latter study confirmed that COD solubilisation from WAS correlated well with the more labour and time intensive degree of disintegration test. Hog manure was found to be more amenable to ultrasonication than WAS, as it took only 3000 kJ/kg TS to cause 15% more solubilisation as compared to 25,000 kJ/kg TS for WAS [11].

For all the tested times in the present study (20, 40, 60, 90, 120, and 180 minutes), COD solubilization slightly increased compared to the untreated sample, being for the longest time applied (180 min) and for the time of 40 min the lowest COD solubilization increase. Ultrasound pretreatment can affect the particle size; some studies establish a relationship between the increase in the sonication time and the particle size concluding that at higher exposure times higher solubilization and lower particle sizes were found, but for very long times of exposure to the ultrasound the opposite effect might be produced owing to the formation of recalcitrant compounds [9]. It has been also reported in the literature that high specific energies may induce the reagglomeration of particles, thereby shifting particle size toward higher diameters, decreasing slightly or keeping constant the solubilization levels [27]. The latter study revealed how the percentage of COD solubilization was maintained around 8% when the specific energy applied increased from 76.5 to 128.9 MJ/kg during the US pretreatment of algal biomass [27].

Other authors found that although sonication disrupted cellular matter providing a higher solubilization than without pretreatment for WAS, the solubilization resulted in soluble nonbiodegradable compounds [10]. The increase in sonication time causes more release of intracellular polymers; these biopolymers released were thought to be the glue that holds bioflocs together [13, 20].

Table 4 shows the hemicellulose, cellulose, and lignin contents of the ultrasound pretreated two-phase OMSWs compared to untreated OMSW. The highest increase in the hemicellulose content (31%) was obtained for the OMSW pretreated during 90 min. The increase in the cellulose content was evident for all the ultrasound pretreatments: 150%, 176%, and 162% for 20, 90, and 180 min, respectively, compared to the untreated two-phase OMSW. In the same way, an increase of 54% in the percentage of cellulose with respect to its initial content in the substrate was observed in the sunflower oil cake after sonication with a specific energy of 24.000 kJ/kg TS [28].

4.3.2 IMPACT OF ULTRASOUND PRETREATMENT ON BIOCHEMICAL METHANE POTENTIAL

The methane yields obtained through BMP after 20 days of digestion for the ultrasound pretreatments selected were 311 + 15, 393 + 14, and 370 + 20 mLCH_4/g VS_{added} for pretreated OMSW during 20, 90, and 180 min, respectively (Figures 1–3), and mLCH_4/g VS_{added} for OMSW without ultrasound pretreatment (Figure 4). The maximum value of methane yield was obtained after a pretreatment time of 90 minutes with a specific energy of 51284 kJ/kg TS and this maximum value was only 5.6% higher than that obtained for untreated OMSW. Higher increments in biogas production and methane yields were reported after sonication of other substrates when compared with untreated samples. For instance, Bougrier et al. [29] showed an increase in the methane yield of WAS from 221 to 334 mLCH_4/g COD_{added} after an ultrasonic pretreatment at 9350 kJ/kg TS, which was more effective than other pretreatments assessed such as ozonation or thermal pretreatment. In the same way, an increase in the methane production of 44% was also reported by Erden and Filibeli [30] for WAS previously sonicated with a specific energy of 9690 kJ/kg TS. Likewise, an improvement of 16% in specific biogas production was also observed after ultrasonic pretreatment of WAS with a high content of polycyclic aromatic hydrocarbons at a specific energy of 11000 kJ/kg TS [16]. Similarly, the methane potential of hog manure increased by 20.7% in comparison with unsonicated manure for a specific energy input of 30000 kJ/kg TS [11], which is lower than that used in the present work for obtaining the maximum methane yield (51284 kJ/kg TS).

In the present study, methane yield slightly increased when the pretreatment time was increased from 20 to 90 minutes and the specific energy consequently increased from 11367 to 51284 kJ/kg TS. A slight decrease in the methane yield was observed for an exposure time of 180 minutes with a specific energy of 106003 kJ/kg TS. The methane yield increase from 20 to 90 minutes may be attributed to the transformation of the particulate part of the substrate to more soluble substances by ultrasonication [28]. When high times of exposition (e.g., 180 min.) are applied the opposite effect can be produced. The organic structures may get more complex owing to the polymeric matter released [13], becoming more difficult to biodegrade. Moreover, high intensive degree times of disintegration test have been found responsible for refractory compound formation and generation of soluble nonbiodegradable compounds which can inhibit methane production [11].

It has been recently demonstrated that the increased solubilization provoked by thermal and ultrasonic pretreatments on mixed-microalgal biomass was not followed by an increased methane production in BMP tests [31]. In the latter study the pretreatments enhanced the transformation of simple sugars to smaller carbon organic acids, especially propionic acid, which results in inhibition of methanogenic microorganisms at certain concentrations [31]. Alzate et al. [32] have recently demonstrated the lack of correlation between the solubilization degree and methane enhancement potential in BMP tests of microalgae mixtures subjected to ultrasound pretreatment. They found no increases in methane productivity with increases in energy inputs at applied energies higher than 10.000 kJ/kg TS.

4.3.3 EFFECT OF ULTRASOUND PRETREATMENT ON THE PROCESS KINETICS

Figures 1, 2, 3, and 4 show the evolution of methane production with time for ultrasonically pretreated two-phase OMSW at 20, 90, and 180 min and untreated OMSW, respectively.

Two different stages were observed for all the cases studied: a first stage during the first 5–7 days of operation followed by an intermediate adaptation period or lag stage and finally a second stage, in which the methane production rate increased gradually to become almost zero at the 20–25 days of digestion.

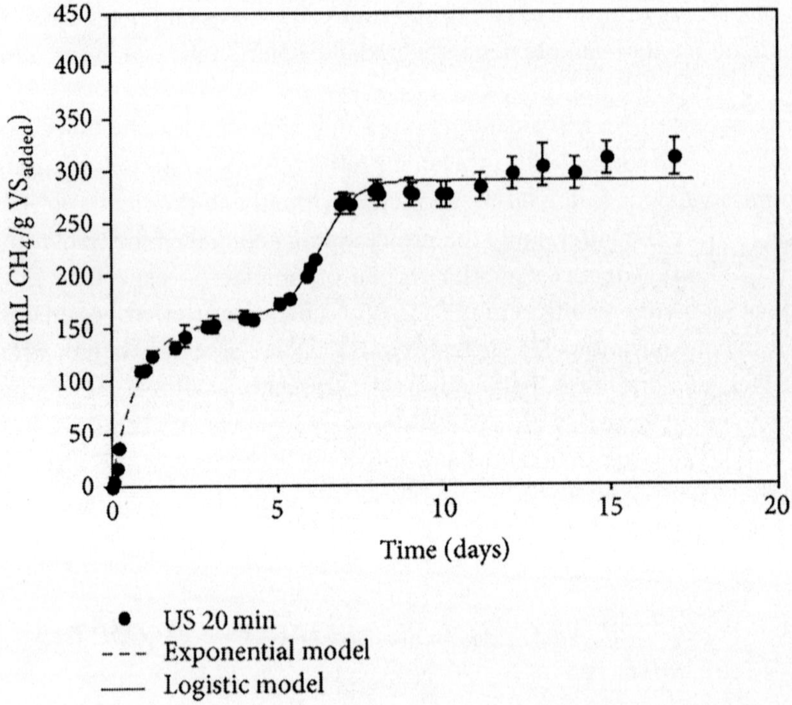

FIGURE 1: Cumulative methane yield, expressed as mL CH_4/g VS_{added}, obtained during the BMP tests carried out with pretreated OMSW with ultrasound during 20 minutes.

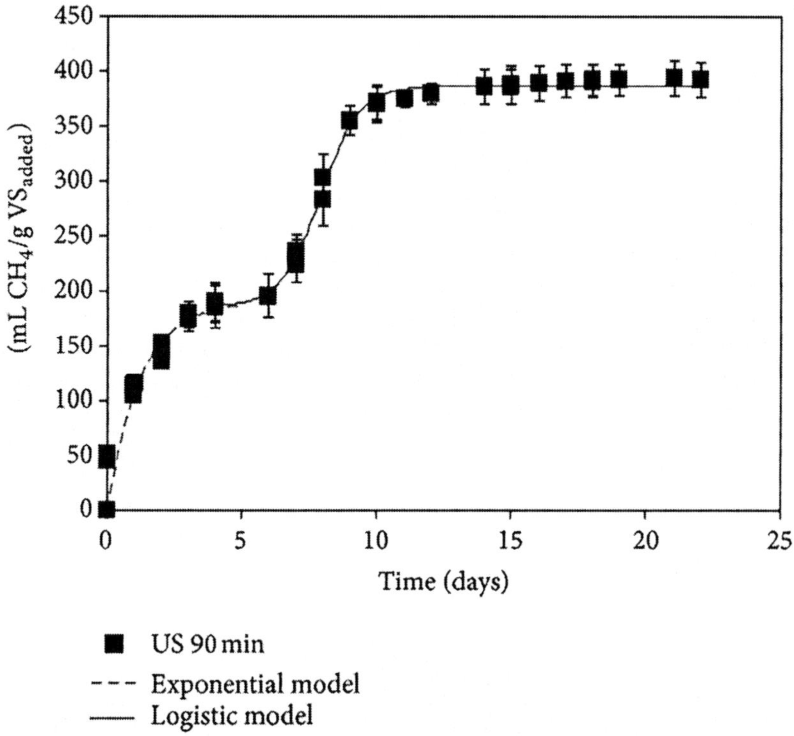

FIGURE 2: Cumulative methane yield, expressed as mL CH_4/g VS_{added}, obtained during the BMP tests carried out with pretreated OMSW with ultrasound during 90 minutes.

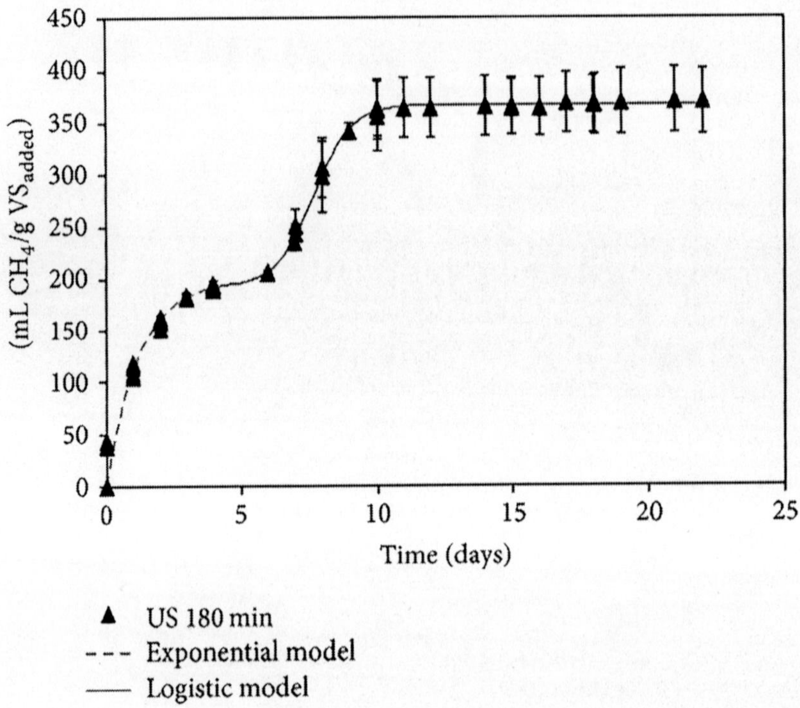

FIGURE 3: Cumulative methane yield, expressed as mL CH_4/g VS_{added}, obtained during the BMP tests carried out with pretreated OMSW with ultrasound during 180 minutes.

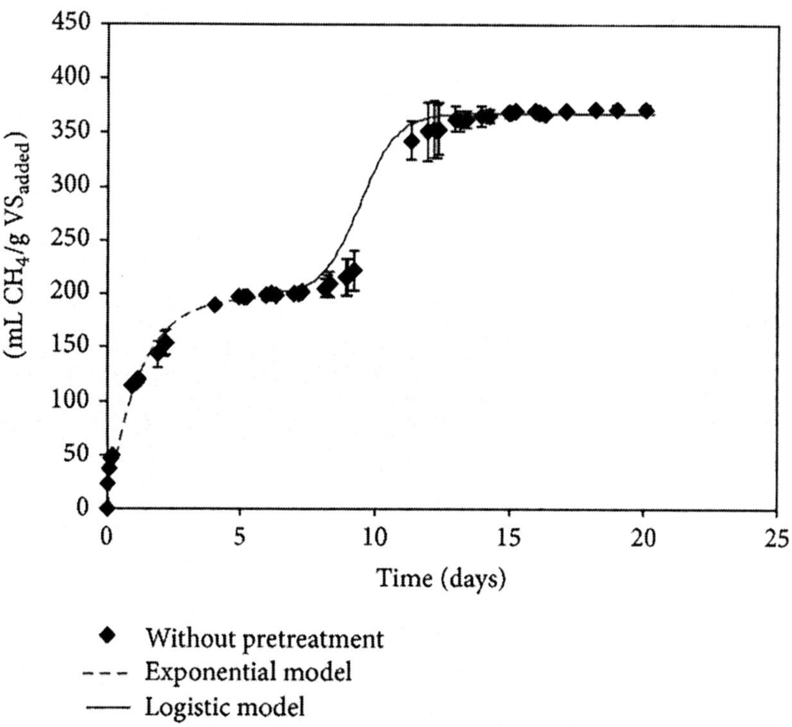

FIGURE 4: Cumulative methane yield, expressed as mL CH_4/g VS_{added}, obtained during the BMP tests carried out with untreated OMSW.

In order to simulate the two stages observed, two different models were used and selected as previously by Rincón et al. [4] with thermally pretreated OMSW: a first-order exponential model for the first stage which is commonly applicable to easily biodegradable substrates [33] and a second sigmoidal or logistic model with its three characteristic phases, that is, lag, exponential increase, and final stabilization step [34].

4.3.3.1 FIRST PHASE: FIRST-ORDER EXPONENTIAL MODEL

The first-order exponential model is given by the following expression:

$$B_1 = B_{max} * [1 - \exp(-K * t)] \qquad (2)$$

where B_1 ($mL CH_4/g\ VS_{added}$) is the cumulative specific methane production, B_{max} ($mL CH_4/g\ VS_{added}$) is the ultimate methane production, K is the specific rate constant or apparent kinetic constant ($days^{-1}$), and (days) is the time.

This model was applied for the first experimental stage of methane production or exponential step (from 0 to 5–7 days) for all the substrates tested. The adjustment by nonlinear regression of the pairs of experimental data (B_1, t) using the Sigmaplot software (version 11.0) allowed the calculation of the parameters K and B_{max} for this first stage of methane production (Table 5). The high values of the R^2 and the low values of the standard error of estimate (S.E.E.) for the cases tested demonstrate the goodness of the fit of experimental data to the model proposed for this first exponential stage.

Table 5 shows the specific rate constants (K) obtained for the first stage of digestion (untreated two-phase OMSW and pretreated two-phase OMSWs at 200 W during 20, 90, and 180 min) with values ranging between 0.82 ± 0.06 and $1.21 + 0.14$ $days^{-1}$. K was significantly higher for the ultrasound pretreatment at 20 min ($K = 1.21 d^{-1}$) than for the other times studied. For the other pretreatment times studied, that is, 90 and 180 min, and for the untreated OMSW, the K values were practically similar ranging between 0.82 and 0.83 d^{-1}. Therefore, the kinetic constant for the ultrasound pretreatment at 20 minutes was 46% higher than those obtained for the pretreated OMSW at 90 and 180 minutes and 48% higher than for untreated OMSW. The highest value of K (1.21 $days^{-1}$) achieved for the pretreated OMSWs during 20 min might be associated with its lower lignin (14.4%) and hemicellulose concentrations (10.9%) after pretreatment compared to the other pretreatment conditions (Table 4). In addition, the values of the kinetic constants obtained in the present research work for the ultrasound pretreated OMSW at 90 and 180 minutes were of the same order of magnitude as those obtained in BMP tests of thermally treated OMSW at 180°C during 180 minutes [4].

During the first stage the ultimate methane production, B_{max}, for the US (20 min) was somewhat lower (158 mL CH_4/g VS_{added}) than those obtained for the other pretreatment times, whose value ranged between 191 mL CH_4/g VS_{added} (US 90 min) and 199 mL CH_4/g VS_{added} (US 180 min). These results might indicate a slight increase of easily degradable compounds after 90 and 180 minutes of pretreatment but still with a high percentage of complex substrates diminishing the degradation rate.

4.3.3.2 SECOND PHASE: SIGMOIDAL OR LOGISTIC MODEL APPLICATION

For the second stage of methane production, that is, between the 5th and 7th days and last day of the operating period, 25th day, the following logistic model (3) was used to estimate process performance [4, 33, 34]:

$$B_2 = B_0 + \frac{P}{[1 + \exp(-4 \cdot R_m \cdot (t - \lambda)/(P + 2))]} \quad (3)$$

where B_2 is the cumulative methane production during the second stage (mL CH_4/g VS_{added}), B_0 is the cumulative methane production at the startup of the second stage (mLCH$_4$/g VS$_{added}$) and should approximately coincide with the value of B_{max} obtained at the end of the first stage, P is the maximum methane production obtained in the second stage (mLCH$_4$/g VS$_{added}$), R_m is the maximum methane production rate (mLCH$_4$/g VS$_{added}$ d), and λ is the lag time (days).

The logistic model assumes the rate of methane production to be proportional to microbial activity [35]. This model has been previously used for estimating the methane production in batch anaerobic digestion experiments of different substrates such as landfill leachate, herbaceous grass materials, and sewage sludge [33–36].

TABLE 5: Kinetic parameters obtained from the exponential model in the BMP tests of untreated OMSW and ultrasound pretreated OMSW at 20, 90, and 180 min.

Time (min)	B_{max} (mLCH$_4$/g VS$_{added}$)	K (days^{-1})	R^2	S.E.E.
Untreated OMSW	197 ± 4	0.82 ± 0.06	0.97	11.3
20	158 ± 7	1.21 ± 0.14	0.94	5.2
90	191 ± 11	0.83 ± 0.17	0.92	12.2
180	199 ±10	0.83 ± 0.16	0.93	7.7

B_{max} is the ultimate methane production; K is the specific rate constant or apparent kinetic constant. Parameters from the nonlinear regression fit: R^2: coefficient of determination; S.E.E.: standard error of estimate.

For the logistic model the maximum methane production obtained in the second stage (P) had the maximum value for the 90-minute pretreatment (200 mLCH$_4$/g VS$_{added}$) followed by the 180-minute pretreatment (174 mLCH$_4$/g VS$_{added}$), untreated OMSW (171 mLCH$_4$/g VS$_{added}$), and pretreatment during 20 minutes (130 mLCH$_4$/g VS$_{added}$) (Table 6). Moreover, comparing the values of the R_m or maximum methane production rates obtained in the logistic model (Table 6) the best pretreatment was the US (90 min). For the ultrasound pretreatment at 90 min the kinetics was the quickest; 70.5 mLCH$_4$/(g VS$_{added}$·day) was produced, a value 12% higher than that obtained for untreated OMSW and 9.5% and 10.3% higher than that obtained at 20 and 180 minutes, respectively.

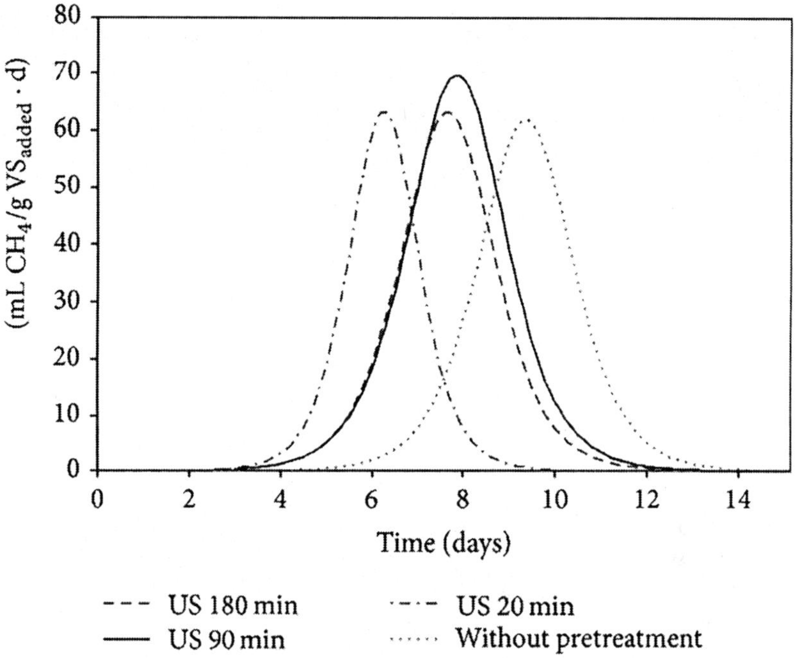

FIGURE 5: Methane production rate, expressed as mL CH4/(g VS d), obtained during the second stage of the BMP tests carried out with untreated OMSW and ultrasound pretreated OMSW during 20, 90, and 180 minutes.

TABLE 6: Kinetic parameters obtained from the logistic model in the BMP tests of untreated OMSW and ultrasound pretreated OMSW at 20, 90, and 180 min.

Time	B_0	P	R_m	L	R^2	S.E.E.
(min)	($mLCH_4/gVS_{added}$)	($mLCH_4/gVS_{added}$)	($mLCH_4/$ $gVS\cdot d$)	(days)		
Untreated OMSW	198 ± 4	171 ± 4	62.7	9.4 ± 0.1	0.99	3.4
20	160 ± 5	130 ± 55	64.4	6.3 ± 0.8	0.96	7.6
90	187 ± 9	200 ± 9	70.5	7.9 ± 0.2	0.99	5.3
180	193 ± 8	174 ± 8	63.9	7.7 ± 0.2	0.99	2.1

B_0 is the cumulative methane production at the startup of the second stage, P is the maximum methane production obtained in the second stage, R_m is the maximum methane production rate, and λ is the lag time. Parameters from the nonlinear regression fit: R^2: coefficient of determination; S.E.E.: standard error of estimate.

The ultrasound pretreatment during 90 minutes most likely promotes the release of more easily biodegradable compounds, which allowed an increase in the R_m and a decrease in the lag period.

The shortest lag phase (λ) was obtained for US (20 min), that is, 6.3 days, while the longest lag phase was achieved for the untreated OMSW, that is, 9.4 days. Long lag phases can lead to the generation of different inhibitor compounds that delay the startup of the second phase in the methane production [34]. The lowest R_m value, that is, 62.7 $mLCH_4/g$ $VS_{added}\cdot d$, was obtained for the pretreatment with the highest lag phase, that is, 9.4 days (untreated OMSW). This value of R_m was very similar to that achieved in BMP tests of OMSW previously treated thermally at 180°C during 180 min [4].

The first derived B_2 with respect to the digestion time gives the evolution of the methane production rate with time during the second stage ($mLCH_4/(gVS\cdot day)$) (Figure 5). The degradation rate of ultrasound pretreated OMSW during 90 minutes was the fastest of the four conditions tested, achieving a maximum methane production rate (R_m) of 70.5 $mLCH_4/(gVS\cdot day)$ after 7.9 days of digestion period. Although the maximum methane production rate for ultrasound pretreated OMSW for 20 minutes was somewhat lower (64.4 $mLCH_4/(gVS\cdot day)$) than that mentioned for pretreated OMSW during 90 minutes, it was achieved at a lower time of 6.3 days. Finally, the methane production rate for untreated OMSW achieved the lowest R_m value and it needed the highest time (9.4 days) to be reached.

Ultrasound pretreatment during 90 minutes gives the most promising results for a fast OMSW degradation process making available a large concentration of soluble and biodegradable components. Ultrasound pretreatment during 180 minutes had opposite effect, indicating a possible recalcitrant compound formation [13, 20].

4.3.4 ENERGY BALANCE

A net balance of the consumed energy in the pretreatment and the produced energy through BMP for the ultrasound pretreated OMSW was found to be negative for all pretreatment times and specific energies studied. The less unfavorable energy balance was observed for the lowest exposure time and specific energy studied, that is, 20 minutes, with a negative energy balance of −1830 kJ/kg TS. For pretreatments of 90 and 180 minutes the energy balance between the consumed energy in the pretreatment (input energy) and produced energy through anaerobic digestion (output energy) was obviously more negative.

A similar negative energy balance has been recently reported in the evaluation of ultrasonic pretreatment combined with anaerobic digestion of mixed-microalgal biomass [31]. After applying thermal, ultrasonic, and alkali pretreatments to raw microalgae biomass to promote the anaerobic digestion efficiency through BMP tests it was observed that only the chemical pretreatment yielded slightly higher energy gains than that of nonpretreatment condition, while the energy balance with the ultrasonic pretreatment gave a negative value of −220 kJ/kg VS using a pretreatment time Of 180 seconds [31]. Houtmeyers et al. [37] reported that the pretreatment of WAS with ultrasounds and microwave both with energy specific of 2100 kJ/kg sludge was economically not feasible although an increase in the biogas production of 27% (microwave pretreated) and 23% (ultrasonic pretreated) was observed with respect to untreated samples. Likewise, a study of the effect of ultrasonic pretreatment on methane production potential from some corn ethanol products (distiller's wet grains, thin stillage, and condensed distiller's solubles) revealed that ultrasonic pretreatment required more energy than was generated by the process in terms of additional biogas production giving a negative energy balance [38]. The efficiency and economic viability of ultrasonication as a pretreatment method for hog manure anaerobic digestion was evaluated at specific energies of 250–30000 kJ/kg TS [11]. Hog manure was found more amenable to ultrasonication than waste activated sludge, as it took only 3000 kJ/kg TS to cause 15% more solubilization as compared to 25000 kJ/kg TS for waste activated sludge. It was noted in this study that biomass cell rupture occurred at specific energy of 500 kJ/kg TS. However, an economic evaluation indicated that only a specific energy of 500 kJ/kg TS was economical, with a net energy output valued at $4.1/ton of dry solids, due to a 28% increase in methane production [11].

4.4 CONCLUSIONS

Ultrasound pretreatment of two-phase OMSW at a power of 200 W (100% amplitude) and a constant frequency of 24 kHz during 20, 90, and 180 minutes increased the COD solubilization of this substrate compared to the untreated sample. The best methane yield obtained through BMP tests, 393 mL CH4/g VS_{added}, was achieved for ultrasound pretreatment during 90 min; this yield was 5.6% higher than that obtained for OMSW without pretreatment. Moreover, taking into account the kinetics of the two stages observed during methane production (exponential and sigmoidal curves), the highest maximum methane production rate was also achieved for ultrasound pretreatment during 90 min. The maximum value of was found for ultrasound pretreatment during 90 min, values 12%, 9.5%, and 10.3% higher than that obtained for untreated OMSW and OMSW pretreated at 20 and 180 min, respectively. A net balance between the consumed energy during the pretreatment and energy production through BMP gave a negative value for all the cases studied.

REFERENCES

1. J. Alba, F. J. Hidalgo, M. A. Ruiz et al., "Elaboración de aceite de oliva virgen," in El Cultivo del Olivo, D. Barranco, R. Fernández-Escobar, and L. Rallo, Eds., pp. 551–588, Mundi-Prensa, Madrid, Spain, 2001.
2. J. A. Alburquerque, J. Gonzálvez, D. García, and J. Cegarra, "Agrochemical characterisation of "alperujo", a solid by-product of the two-phase centrifugation method for olive oil extraction," Bioresource Technology, vol. 91, no. 2, pp. 195–200, 2004.
3. R. Borja, B. Rincón, F. Raposo, J. Alba, and A. Martín, "A study of anaerobic digestibility of two-phases olive mill solid waste (OMSW) at mesophilic temperature," Process Biochemistry, vol. 38, no. 5, pp. 733–742, 2002.
4. B. Rincón, L. Bujalance, F. G. Fermoso, A. Martín, and R. Borja, "Biochemical methane potential of two-phase olive mill solid waste: influence of thermal pretreatment on the process kinetics," Bioresource Technology, vol. 140, pp. 249–255, 2013.
5. B. Rincón, L. Travieso, E. Sánchez et al., "The effect of organic loading rate on the anaerobic digestion of two-phase olive mill solid residue derived from fruits with low ripening index," Journal of Chemical Technology and Biotechnology, vol. 82, no. 3, pp. 259–266, 2007.
6. B. Rincón, R. Borja, J. M. González, M. C. Portillo, and C. Sáiz-Jiménez, "Influence of organic loading rate and hydraulic retention time on the performance, stability and microbial communities of one-stage anaerobic digestion of two-phase olive mill solid residue," Biochemical Engineering Journal, vol. 40, no. 2, pp. 253–261, 2008.
7. L. Appels, J. Lauwers, J. Degrve et al., "Anaerobic digestion in global bio-energy production: potential and research challenges," Renewable and Sustainable Energy Reviews, vol. 15, no. 9, pp. 4295–4301, 2011.
8. T. Shimizu, K. Kudo, and Y. Nasu, "Anaerobic waste-activated sludge digestion—a bioconversion mechanism and kinetic model," Biotechnology and Bioengineering, vol. 41, no. 11, pp. 1082–1091, 1993.
9. A. Tiehm, K. Nickel, and U. Neis, "The use of ultrasound to accelerate the anaerobic digestion of sewage sludge," Water Science and Technology, vol. 36, no. 11, pp. 121–128, 1997.
10. M. Saha, C. Eskicioglu, and J. Marin, "Microwave, ultrasonic and chemo-mechanical pretreatments for enhancing methane potential of pulp mill wastewater treatment sludge," Bioresource Technology, vol. 102, no. 17, pp. 7815–7826, 2011.
11. E. Elbeshbishy, S. Aldin, H. Hafez, G. Nakhla, and M. Ray, "Impact of ultrasonication of hog manure on anaerobic digestability," Ultrasonics Sonochemistry, vol. 18, no. 1, pp. 164–171, 2011.
12. A. Elliott and T. Mahmood, "Pretreatment technologies for advancing anaerobic digestion of pulp and paper biotreatment residues," Water Research, vol. 41, no. 19, pp. 4273–4286, 2007.
13. S. Pilli, P. Bhunia, S. Yan, R. J. LeBlanc, R. D. Tyagi, and R. Y. Surampalli, "Ultrasonic pretreatment of sludge: a review," Ultrasonics Sonochemistry, vol. 18, no. 1, pp. 1–18, 2011.

14. H. Carrère, C. Dumas, A. Battimelli et al., "Pretreatment methods to improve sludge anaerobic degradability: a review," Journal of Hazardous Materials, vol. 183, no. 1–3, pp. 1–15, 2010.
15. A. Tiehm, K. Nickel, M. Zellhorn, and U. Neis, "Ultrasonic waste activated sludge disintegration for improving anaerobic stabilization," Water Research, vol. 35, no. 8, pp. 2003–2009, 2001.
16. C. Bougrier, H. Carrère, and J. P. Delgenès, "Solubilisation of waste-activated sludge by ultrasonic treatment," Chemical Engineering Journal, vol. 106, no. 2, pp. 163–169, 2005.
17. F. Hogan, S. Mormede, P. Clark, and M. Crane, "Ultrasonic sludge treatment for enhanced anaerobic digestion," Water Science and Technology, vol. 50, no. 9, pp. 25–32, 2004.
18. E. Gonze, S. Pillot, E. Valette, Y. Gonthier, and A. Bernis, "Ultrasonic treatment of an aerobic activated sludge in a batch reactor," Chemical Engineering and Processing: Process Intensification, vol. 42, no. 12, pp. 965–975, 2003.
19. T. B. El-Hadj, J. Dosta, R. Márquez-Serrano, and J. Mata-Álvarez, "Effect of ultrasound pretreatment in mesophilic and thermophilic anaerobic digestion with emphasis on naphthalene and pyrene removal," Water Research, vol. 41, no. 1, pp. 87–94, 2007.
20. L. L. Zhang, X. X. Feng, N. W. Zhu, and J. M. Chen, "Role of extracellular protein in the formation and stability of aerobic granules," Enzyme and Microbial Technology, vol. 41, no. 5, pp. 551–557, 2007.
21. U. Nels, K. Nickel, and A. Tiehm, "Enhancement of anaerobic sludge digestion by ultrasonic disintegration," Water Science and Technology, vol. 42, no. 9, pp. 73–80, 2000.
22. APHA-AWWA-WPCF, Standard Methods for the Examination of Water and Wastewater, APHA-AWWA-WPCF, Washington, DC, USA, 20th edition, 1998.
23. P. J. Van Soest, J. B. Robertson, and B. A. Lewis, "Methods for dietary fiber, neutral detergent fiber, and nonstarch polysaccharides in relation to animal nutrition," Journal of Dairy Science, vol. 74, no. 10, pp. 3583–3597, 1991.
24. J. Kim, C. Park, T. Kim et al., "Effects of various pretreatments for enhanced anaerobic digestion with waste activated sludge," Journal of Bioscience and Bioengineering, vol. 95, no. 3, pp. 271–275, 2003.
25. H. Carrère, B. Sialve, and N. Bernet, "Improving pig manure conversion into biogas by thermal and thermo-chemical pretreatments," Bioresource Technology, vol. 100, no. 15, pp. 3690–3694, 2009.
26. Q. Wang, M. Kuninobu, K. Kakimoto, H. I. Ogawa, and Y. Kato, "Upgrading of anaerobic digestion of waste activated sludge by ultrasonic pretreatment," Bioresource Technology, vol. 68, no. 3, pp. 309–313, 1999.
27. C. González-Fernández, B. Sialve, N. Bernet, and J. P. Steyer, "Comparison of ultrasound and thermal pretreatment of Scenedesmus biomass on methane production," Bioresource Technology, vol. 110, pp. 610–616, 2012.
28. V. Fernández-Cegrí, M. A. de la Rubia, F. Raposo, and R. Borja, "Impact of ultrasonic pretreatment under different operational conditions on the mesophilic anaerobic digestion of sunflower oil cake in batch mode," Ultrasonics Sonochemistry, vol. 19, no. 5, pp. 1003–1010, 2012.

29. C. Bougrier, C. Albasi, J. P. Delgenès, and H. Carrère, "Effect of ultrasonic, thermal and ozone pre-treatments on waste activated sludge solubilisation and anaerobic biodegradability," Chemical Engineering and Processing, vol. 45, no. 8, pp. 711–718, 2006.
30. G. Erden and A. Filibeli, "Ultrasonic pre-treatment of biological sludge: consequences for disintegration, anaerobic biodegradability, and filterability," Journal of Chemical Technology and Biotechnology, vol. 85, no. 1, pp. 145–150, 2010.
31. S. Cho, S. Park, J. Seon, J. Yu, and T. Lee, "Evaluation of thermal, ultrasonic and alkali pretreatments on mixed-microalgal biomass to enhance anaerobic methane production," Bioresource Technology, vol. 143, pp. 330–336, 2013.
32. M. E. Alzate, R. Muñoz, F. Rogalla, F. Fdz-Polanco, and S. I. Pérez-Elvira, "Biochemical methane potential of microalgae: influence of substrate to inoculum ratio, biomass concentration and pretreatment," Bioresource Technology, vol. 123, pp. 488–494, 2012.
33. L. Li, X. Kong, F. Yang, D. Li, Z. Yuan, and Y. Sun, "Biogas production potential and kinetics of microwave and conventional thermal pretreatment of grass," Applied Biochemistry and Biotechnology, vol. 166, no. 5, pp. 1183–1191, 2012.
34. A. Donoso-Bravo, S. I. Pérez-Elvira, and F. Fdz-Polanco, "Application of simplified models for anaerobic biodegradability tests. Evaluation of pre-treatment processes," Chemical Engineering Journal, vol. 160, no. 2, pp. 607–614, 2010.
35. L. Altaş, "Inhibitory effect of heavy metals on methane-producing anaerobic granular sludge," Journal of Hazardous Materials, vol. 162, no. 2-3, pp. 1551–1556, 2009.
36. S. Pommier, D. Chenu, M. Quintard, and X. Lefebvre, "A logistic model for the prediction of the influence of water on the solid waste methanization in landfills," Biotechnology and Bioengineering, vol. 97, no. 3, pp. 473–482, 2007.
37. S. Houtmeyers, L. Appels, J. Degrève, J. V. Impe, and R. Dewil, "Comparing the influence of ultrasonic and microwave pre-treatment on the solubilization and semi-continuous digestion of waste activated sludge," in Proceedings of the Anaerobic Digestion Conference, Paper SPC21, Santiago de Compostela, Spain, June 2013.
38. W. Wu-Haan, R. T. Burns, L. B. Moody, C. J. Hearn, and D. Grewell, "Effect of ultrasonic pretreatment on methane production potential from corn ethanol coproducts," Transactions of the ASABE, vol. 53, no. 3, pp. 883–890, 2010.

PART III

ENERGY RECOVERY

CHAPTER 5

Solid Waste as Renewable Source of Energy: Current and Future Possibility in Algeria

BOUKELIA TAQIY EDDINE AND MECIBAH MED SALAH

5.1 INTRODUCTION

In order to use the enormous source of renewable energies, Algeria has created a green momentum by launching an ambitious program to develop renewable energies (RES) and promote energy efficiency. This program leans on a strategy focused on developing and expanding the use of inexhaustible resources, such as solar, biomass, geothermal, wind, and hydropower, energies in order to diversify energy sources and prepares Algeria for tomorrow.

The program consists of installing up to 22,000 MW of power-generating capacity from renewable sources between 2011 and 2030, of which 12,000 MW will be intended to meet the domestic electricity demand and 10,000 MW destined for export [1]. This last option depends on the availability of a demand that is ensured on the long term by reliable partners as well as on attractive external funding. In this program, it is expected that about 40% of electricity produced for domestic consumption will be from renewable energy sources by 2030.

Solid waste is one of most important sources of biomass potential in Algeria, which is a by-product from human activities, and is characterized by the negative impacts that may affect man and the environment when disposed in an inappropriate way without treatment. Due to the continuously increasing amount of solid waste generated, particularly in capitals and major urban centers, the challenge for the governments is to reduce the waste's harmful impacts to both health and the environment. This paper is an investigation on the possibility to use solid waste as a source of bioenergy in Algeria.

5.2 COUNTRY PROFILE

Algeria, situated in the center of North Africa between the 35° to 38° latitude north and 8° to 12° longitude east, has an area of 2,381,741 km^2[2,3]. In the west, Algeria borders with Morocco, Mauritania, and Occidental Sahara; in the southwest, with Mali; in the east, with Tunisia and Libya; and in the southeast, with Niger (Figure 1). The geographic location of Algeria signifies that it is in a position to play an important strategic role in the implementation of renewable energy technology in the north of Africa.

The climate is transitional between maritime (north) and semi-arid to arid (middle and south). The mean annual precipitation varies from 500 mm (in the north) to 150 mm (in the south). The average annual temperature is around 12°C.

Algeria has a strongly growing population, with 36,275,358 inhabitants in 2011 according to Population Reference Bureau [4]. In the last 25 years, it has almost doubled. Though we notice a slowing down in the 1990s, the last statistics (sizable increase in marriage rate and in fertility rate) indicate that it is a short-term phenomenon, and the population growth is taking a turn towards fast growth. Algeria is characterized by a young and growing population and a fast urbanization. This situation puts certainly a lot of pressure on the energy, food supply, and even on the environment by increasing the generation of waste and residues.

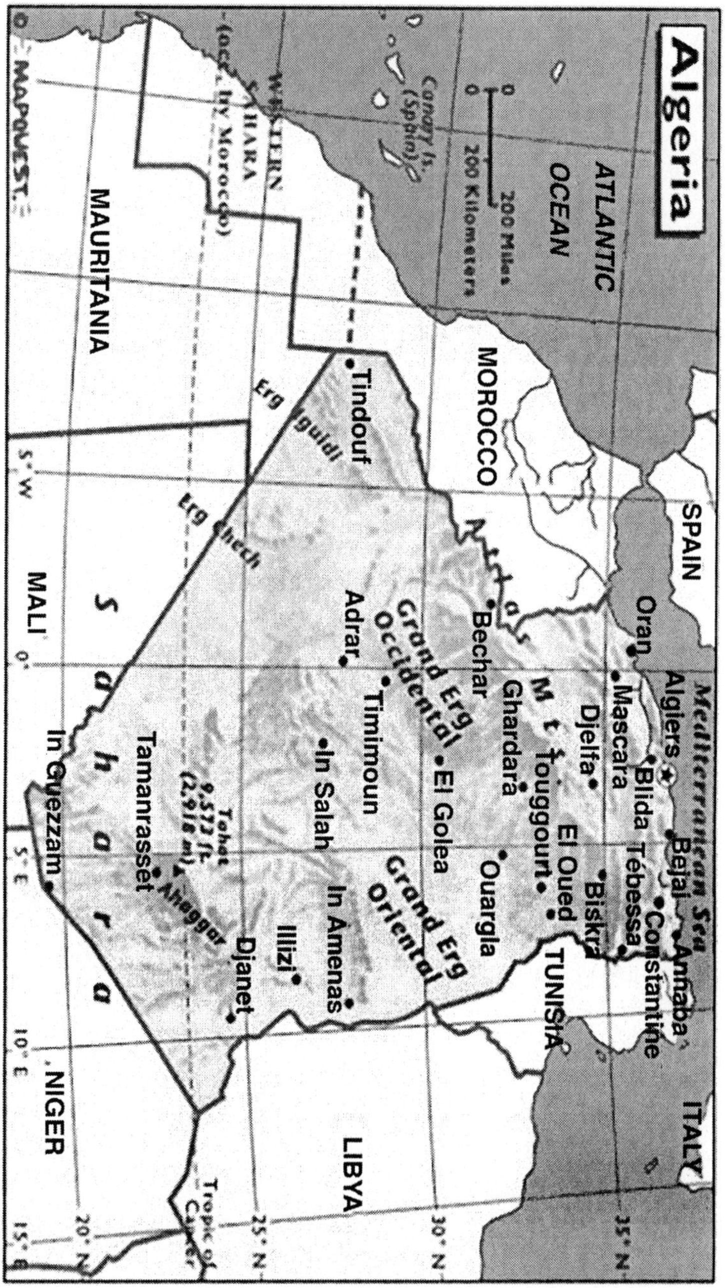

FIGURE 1: Map of Algeria.

Algeria plays a very important role in the world energy markets, both as a significant hydrocarbon producer and as an exporter, as well as a key participant in the renewable energy market. According to the 2011 BP Statistical Energy Survey, in 2010, Algeria had proved a natural gas reserve of 4.5 trillion m^3 and natural gas production of 80.41 billion m^3 with consumption of 28.87 billion m^3. Algeria had proved an oil reserve of 12.2 billion barrels at the end of 2010 and produced an average of 1,809 thousand barrels of crude oil per day, according to the same survey; Algeria consumed an average of 327.03 thousand barrels a day of oil in 2010 [2,3,5].

5.3 PROMOTION OF RENEWABLE ENERGIES IN ALGERIA

Algeria's location has several advantages for extensive use of most of the RES, in which it has very important potential of renewable energies including thermal solar (169,440 TWh/year), photovoltaic solar (13.9 TWh/year), and wind energy (35 TWh/year).

5.4 SOLAR ENERGY

Fortunately, Algeria has enormous potential of solar energy. More than 2,000,000 km^2 receives a yearly sunshine exposure equivalent to 2,500 KWh/m^2. The mean yearly sunshine duration varies from a low of 2,650 h on the coastal line to 3,500 h in the south.

In addition, as shown in Table 1, the potential of daily solar energy is important. It varies from a low average of 4.66 kWh/m^2 in the north to a mean value of 7.26 kWh/m^2 in the south.

Photovoltaic solar energy projects in Algeria: With a potential of 13.9 TWh/year, the government plans launching several solar photovoltaic projects with a total capacity of 800 MW by 2020. Other projects with an annual capacity of 200 MW are to be achieved over the 2021 to 2030 period.

Concentrating solar thermal energy: Pilot projects for the construction of two solar power plants with a storage total capacity of about 150 MW each will be launched during the 2011 to 2013 period. These will be in addition to the hybrid power plant (solar-gas) project of Hassi R'Mel which was built for 130 MW of gas and 25 MW of thermal solar energy with the parabola system of the giant mirrors on a surface of approximately 180,000 m^2.

Four solar thermal power plants with a total capacity of about 1,200 MW are to be constructed over the period of 2016 to 2020. The program of 2021to 2030 provides for the installation of an annual capacity of 500 MW until 2023, then 600 MW per year until 2030.

TABLE 1: Regional daily solar energy and sunshine duration in Algeria [3]

Parameters	Region		
	Coastal line	High plateaux	Sahara
Area (km^2)	95,271	238,174	2,048,296
Mean daily sunshine duration (h)	7.26	8.22	9.59
Solar daily energy density (kWh/m^2)	4.66	5.21	7.26
Potential daily energy (1012 Wh)	443.96	1,240.89	14,870.63

5.5 WIND ENERGY

Wind is another renewable source that is very promising with a potential of 35 TWh/year. The wind map of Figure 2, established by the Centre of Renewable Energies Development (CDER) and the Ministry of Energy and Mines (MEM), shows that 50% of the country surface presents a considerable average speed of the wind. The map also shows that the south-western region experiences high wind speeds for a significant fraction of the year. The Algerian RES program plans at first, in the period of 2011 to 2013, the installation of the first wind farm with a power of 10 MW in Adrar. Between 2014 and 2015, two wind farms with a capacity of 20 MW each are to be developed. Studies will be led to detect suitable sites to realize the other projects during the period of 2016 to 2030 for a power of about 1,700 MW.

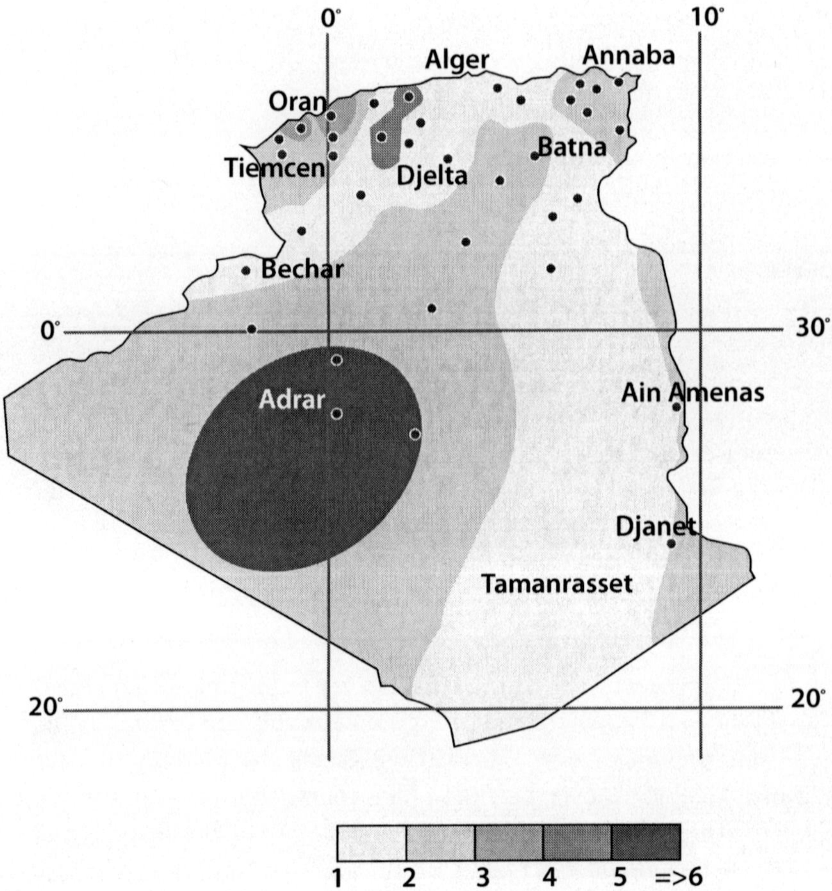

FIGURE 2: Wind chart of Algeria.

5.6 GEOTHERMAL ENERGY

The Algerian potential of geothermal energy is estimated at 460 GWh/year [7]. More than 200 geothermal sources were counted by the CDER [7] and are recorded, of which one-third of the temperatures are superior to 45°C and where the highest temperatures registered are 98°C and 118°C in Hamam El Maskhoutin and Biskra, respectively, situated in the western part of the country. So far, the applications are limited to agricultural (heating of greenhouses, aquaculture), space heating, sanitary, and balneotherapy.

5.7 HYDROELECTRICITY

The overall flows falling over the Algerian territory are important and estimated to be 65 billion m^3 but of little benefit to the country due to the following reasons: restrained rainfall days, concentration on limited areas, high evaporation, and quick evacuation to the sea.

Schematically, the surface resources decrease from the North to the South. Currently, the evaluated useful and renewable energies are about 25 billion cubic meters, of which approximately two-thirds is for the surface resources. Hydraulic electricity represented, with 265GWh in 2003, barely 1% of the total electricity production.

5.8 BIOMASS POTENTIAL

The biomass potentially offers great promises with a bearing of 3.7 millions of tons of oil equivalent (TEP) coming from forests and 1.33 million of TEP per year coming from agricultural and urban wastes; however, this potential is not enhanced and consumed yet [2,6].

Regulations from the MEM which supports the use of biomass from energy crops rapidly caused an increase consumption of biomass, and in the interest in connecting the agriculture and energy sectors, this is seen as a first step in stimulating the use of biomass in Algeria much faster.

5.9 SOLID WASTE GENERATION IN ALGERIA

In this paper, classifications of solid wastes have been proposed according to its origin into three types: municipal solid waste (MSW), industrial solid waste (ISW), and healthcare solid waste (HW).

According to the National Cadastre for Generation of Solid Waste in Algeria, the quantity of MSW generated in Algeria is estimated at 10.3 million tons/year (household and similar waste). The overall generation of ISW, including non-hazardous and inert industrial waste, is 2,547,000 tons/year with a stock quantity of 4,483,500 tons. The hazardous waste generated amounts to 325,100 tons/year. The quantities of waste in stock and awaiting a disposal solution amount to 2,008,500 tons. Healthcare waste reaches to 125,000 tons/year according to the same source.

5.10 MUNICIPAL SOLID WASTE

MSW is generally defined as waste collected by municipalities or other local authorities. It includes mainly household (domestic waste), commercial, and institutional wastes (generated from shops and institutions). These wastes are generally in solid or semi-solid form. It can be classified as biodegradable waste that includes food and kitchen waste, green waste, and paper (can also be recycled); recyclable materials such as paper, glass, bottles, cans, metals, certain plastics, etc.; inert waste such as construction and demolition wastes, dirt, rocks, and debris; composite waste which includes waste clothing, tetra packs, and waste plastics such as toys; domestic hazardous waste (also called 'household hazardous waste'); and toxic waste like medication, e-waste, paints, chemicals, light bulbs, fluorescent tubes, spray cans, fertilizer and pesticide containers, batteries, and shoe polish.

According to the National Waste Agency (AND), Algeria produces 10.3 million tons of MSW each year or 28,219 tons per day, with a collection coverage of 85% in urban areas and 60% in rural areas, and a rate of 0.9 kg/inhabitant/day for urban zones and 0.6 kg/inhabitant/day for rural zones. In the capital (Algiers), the production is close to 1.2 kg/inhabitant/day [8].

The composition of MSW is closely related to the level of economic development and lifestyle of the residents. In different districts of the same city, the composition of MSW will be different. In general, the composition of MSW in Algeria with six major categories of waste was identified: organic matter, paper-cardboard, plastics, glass, metals, and others (Table 2).

TABLE 2: Waste composition category [17]

Waste category	Waste components
Organic matter	Waste from foodstuff such as food and vegetable refuse, fruit skin, stem of green, corncob, leaves, grass, and manure
Paper	Paper, paper bags, cardboard, corrugated board, box board, newsprint, magazines, tissue, office paper, and mixed paper (all papers that do not fit into other categories)
Plastic	Any material and products made of plastics such as wrapping film, plastic bag, polythene, plastic bottle, plastic hose, and plastic string
Glass	Any material and products made of glass such as bottles, glassware, light bulb, and ceramics
Metal	Ferrous and non-ferrous metal such as tin can, wire, fence, knife, bottle cover, aluminum can and other aluminum materials, foil, ware and bi-metal
Others	Materials from leather, rubber, textile, wood, and others such yard waste, tires, batteries, large appliances, nappies/sanitary products, medical waste, etc.

Organic matter was the predominant category and represented 62% of waste collected. The other categories were represented as follows: paper-cardboard (9%), plastic (12%), glass (1%), metals (2%), and others (14%) (Figure 3). Demolition and construction wastes were not taken into account because they are disposed in uncontrolled open-air sites. The high consumption of fruits and vegetables by the city's inhabitants could explain the preponderance of organic matter in Algeria's waste.

5.11 INDUSTRIAL SOLID WASTE

According to the National Cadastre for Industrial and Special Wastes prepared in 2007, the overall generation of industrial waste, including non-hazardous and inert industrial waste, is 2,547,000 tons per year with a stock quantity of 4,483,500 tons. This type of waste is generated from the following:

- steel, metallurgical, mechanical, and electrical industries, which are the predominant sectors (50%);
- building materials, ceramics, and glass industries (50%);
- chemicals, rubber, and plastic industries (2%);
- food processing, tobacco, and match industries (29%);
- Textiles, hosiery, and confection industries (10%);
- leather and shoes industries (1%); and
- wood, paper, printing industries (3%).

The hazardous waste which includes waste oil, waste solvents, ash, cinder, and other wastes with hazardous nature (such as flammability, explosiveness, and causticity) generated amounts to 325,100 tons/year. The quantities of waste in stock and awaiting a disposal solution amount to 2,008,500 tons, which are generated by four principal sectors: hydrocarbons (34%), chemistry, rubber and plastic (23%), metallurgy (16%), and mines (13%). Compared to textile (4%) as well as paper and cellulose cement and drifts, food and mechanics produce less than 2%.

Table 3 shows that the eastern regions hold the palm for the production of ISW in Algeria, with the wilayas of Annaba and Skikda which are characterized by a high proportion of waste generated and in stock (the petrochemical, transportation, and hydrocarbons industries of these regions). The western region is in the second position, because the industrial area of Arzew is the largest generator of waste with 65,760 T/year only for its refinery, followed by the industrial area of Ghazaouet with 18,500 T/year. The central region is characterized by the high production of lead waste (manufacture of battery and refinery) [11].

5.12 HEALTHCARE WASTE

These wastes include materials like plastic syringes, animal tissues, bandages, cloths, etc. This type of waste results from the treatment, diagnosis, or immunization of humans and/or animals at hospitals, veterinary and health-related research facilities, and medical laboratories. HW contains infectious waste, toxic chemicals, and heavy metals, and may contain substances that are genotoxic or radioactive. HW reach 125,000 tons/year, of which 53.6% is general waste, 17.6% is infectious waste, 23.2% is toxic waste, and 5.6% is special waste, with waste generation rate 0.7 to 1.22 kg/bed/day, in which 75% to 90% is non-clinical waste and 10% to 25% is clinical waste [13,14].

5.13 WASTE MANAGEMENT SITUATION IN ALGERIA

During the past decades, environmentally sound waste management was recognized by most countries as an issue of major concern. Waste management is an important factor in ensuring both human health and environmental protection [15].

5.14 ACTORS OF WASTE MANAGEMENT SERVICES

Policy and planning: The Ministry of Land Planning and the Environment (MATE) is primarily responsible for national policy environment.

Implementation and operation: AND has the mission to support the local communities in SWM and to promote activities linked to integrated waste management.

Practice of waste management:

1. Municipalities are fully responsibility for the management and control of municipal solid waste.
2. The Ministry of the Interior and Local Communities is for financial and logistical support to the municipalities.

Control and regulatory implementation: The Directorate of Environment of each wilaya (governorate) controls and regulates the implementation of the management services.

Staff training: The National Conservatory for Environmental Training does the staff training.

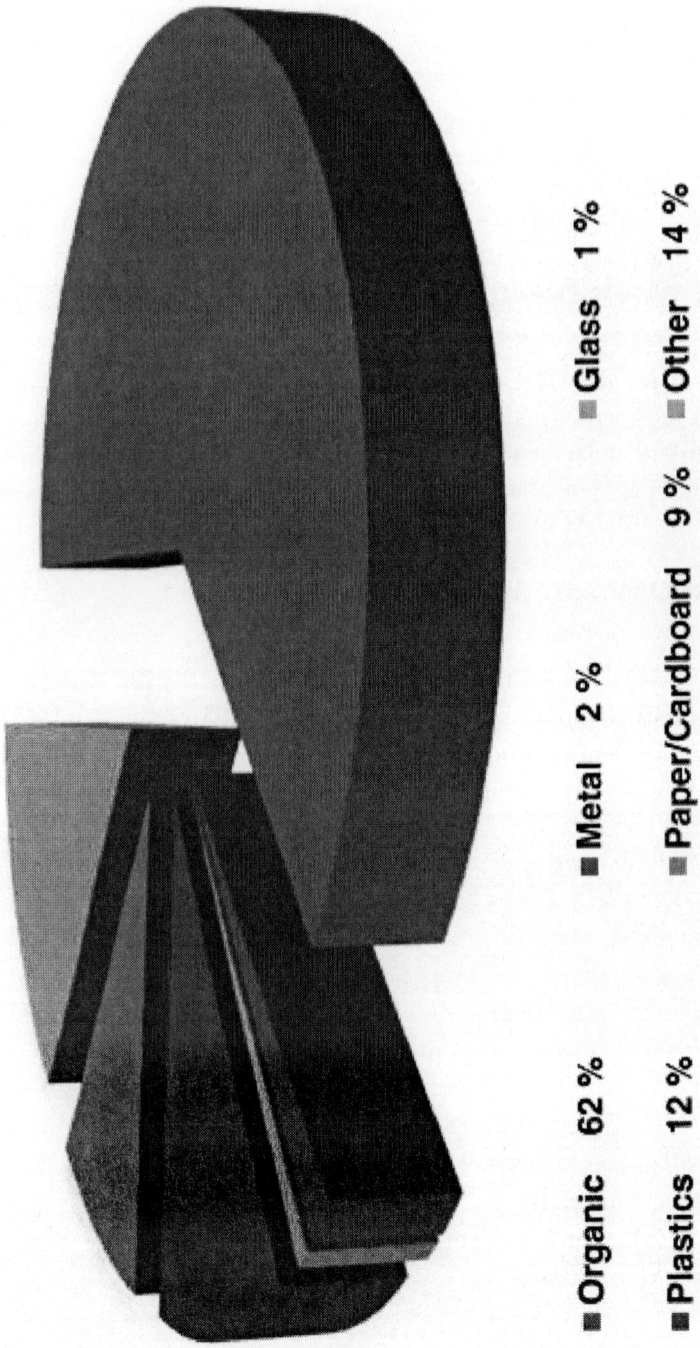

FIGURE 3: MSW composition in Algeria [10].

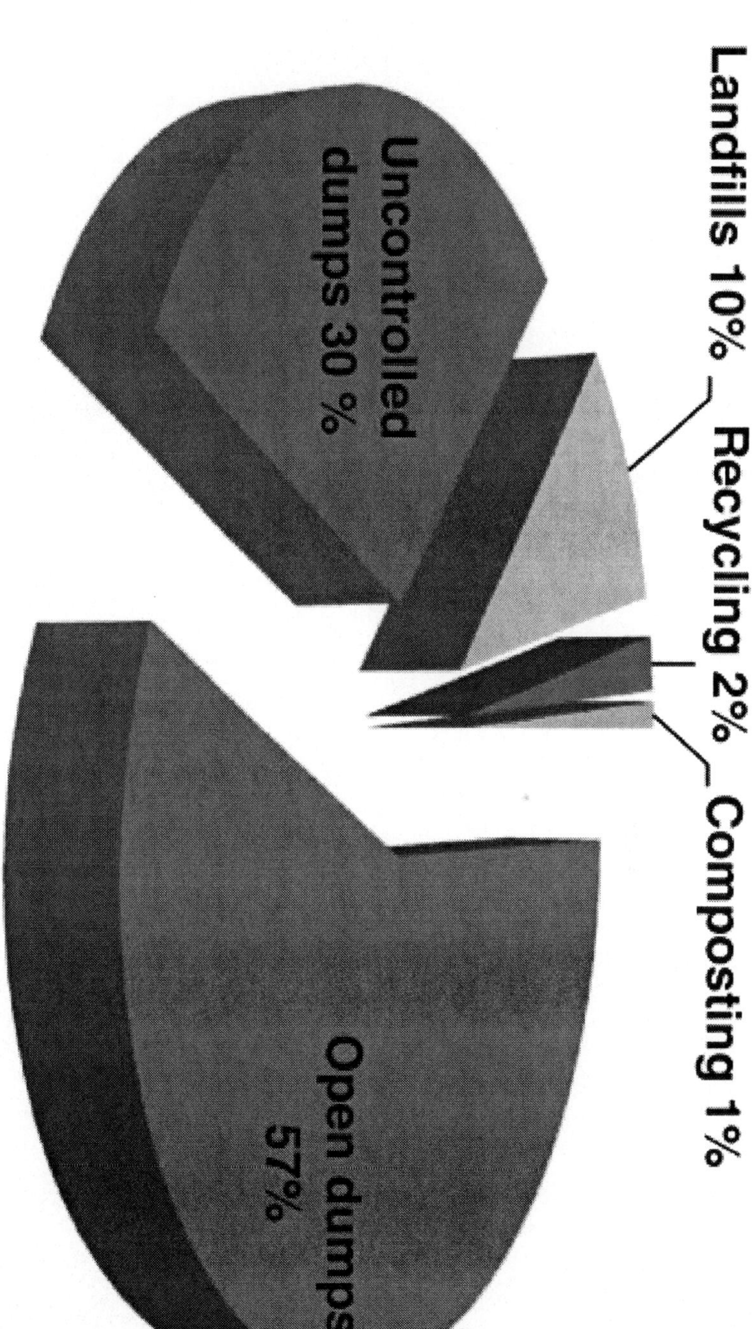

FIGURE 4: Methods of waste disposal in Algeria.

TABLE 3: Production of hazardous industrial waste in Algeria [12]

The area/production site	Production of HIW		Stock of HIW	
	The quantity (tons/year)	%	The quantity (tons/year)	%
The Eastern region	145,000	45	1,100,800	54.6
The Western region	98,500	30	521,800	25.6
The Central region	77,007	24	378,000	19.5
Southeast and Southwest regions	4,500	1.4	7,900	0.3
Total	325,007	100	2,008,500	100

5.15 POLICY AND PLANNING

Municipal Solid Waste Management National Program (PROGDEM): Launched in 2001, it has already made the development of many SWM projects (municipality master schemes, landfills, etc.) possible.

Industrial and Special Waste Management National Program: This program aims at the control and disposal of special industrial waste and potentially infectious healthcare waste.

5.16 SOLID WASTE MANAGEMENT

In general, elimination is the solution applied to 97% of waste produced in Algeria. Wastes are disposed in open dumps (57%), burned in the open air in public dumps or municipal uncontrolled ones (30%), and controlled dumps and landfill (10%) (Figure 4). On the other side, the quantities destined for recovery are too low: only 2% for recycling and 1% for composting [8].

Open dump mode: In Algeria, the elimination of household and similar wastes through the implementation of open and uncontrolled dumps is the most common mode used, with a rate of 87%. According to an investigation by the Office of Ministry of Land Planning and the Environment, over 3,130 open dumps have been identified in the country with an area of approximately 4,552.5 ha [8]. The majority of these dumps are characterized by almost similar geographical locations. They are located along rivers, roads or agriculture lands. The other common point is that most of these dumps are almost saturated and cannot practically receive waste.

We can use the open dump of Oued Smar (Figure 5) as an actual example; it is located 13 km from the center of Algiers and was established in 1978 on an initial area of 10 ha. It now covers 32 ha and receives more than 700 trucks arriving from 56 municipalities, or 2,200 tons/day of MSW and more than 450 tons/day of rubble and fill. Table 4 provides a summary of the amount of waste placed during the period from 1978 to 2007. It went 15,000 tons during the opening year of discharge to more than 350,000 tons at the end of 2007 [8,16].

TABLE 4: Quantities of waste received by the open dumps of Oued Smar (1978 to 2007) [8]

Year	1978	1980	1985	1990	1995	2000	2005	2007
Quantity (tons)	15,332	21,221	47,826	107,787	242,926	336,439	375,263	375,263

Landfill mode: Since 2001, the Algerian government has chosen to eliminate the municipal solid waste by landfill technique, which is a waste storage underground. One of the objectives of PROGDEM is to abandon the traditional mode of disposing waste by open dumps. Following the launching of PROGDEM, 65 landfills were recorded during the period from 2001 to 2005; 16 were completed, 28 under construction, and 21 during the study phase. In the end of 2007, this number has increased due to the results of pilot projects including that of Ouled Fayet in Algiers (Figure 6). It increased to 80 projects, 20 completed, 34 under construction, and 26 in study, or 15 new projects. First, the wilayas concerned are Skikda, El-Tarf, Annaba, Guelma, Souk Ahras, Batna, Tebessa, Media, Tizi-Ouzou, Setif, Biskra, Algiers, M'Sila, Ouargla, Blida, Djelfa, Jijel, Bejaia, and Chelf. In 2010, this number increased to 100 landfills, much of which was nearly completed, according to a communication of same source.

FIGURE 5: The open dump of Oued Smar [8].

FIGURE 6: Landfill of Ouled Fayet [16].

The main advantages of this technology are the following:

- a universal solution that provides ultimate waste disposal;
- relatively low cost and easy to implement to other waste management technology; and
- can derive landfill biogas as a byproduct for household and industrial uses.

However, this technology also has several disadvantages such as landfills requiring a large surface area and pollution problems, including ground water pollution, air pollution, and soil contamination.

The landfill of Ouled Fayet is part of the new policy of integrated waste management which included the transformation of some dumps into landfills. It serves over 34 towns of the wilayas of Algiers and Tipaza. The amount of landfill waste is 864 tons/day in 2005 against 72 tons/day when it opened in 2001. Table 5 shows the evolution of the amount of landfill waste since October 22, 2002.

TABLE 5: Evolution of the amount of landfill waste at the site of Ouled Fayet [8]

Period	Number of days	Number of trips	Tonnage (tons)
22/10/02 to 18/07/03	270	19,588	86,780
19/07/03 to 31/03/04	257	14,034	54,433
06/05/04 to 31/05/06	756	115,086	583,014
22/08/07 to 31/10/07	71	13,625	42,178

Composting mode: Composting is a biological method for recovering organic material in solid waste. Composting represents only 1% of all waste produced in Algeria. The only experiments are those of the wilayas of Blida, Algiers, Tlemcen, and Tizi-Ouzou. The main benefit of this technology is that it converts decomposable organic materials into organic fertilizers.

We can give the example of composting station in the city of Blida; this station was put into service in 1989 and rehabilitated during the period from 1992 to 1996 and returned to service in 1996. It spreads over an area of 3.7 ha for a nominal capacity of 100 tons per shift for 8 h and for a production of 40 tons of compost.

Recovery and recycling mode: Depending on the services of the MATE, Algeria has the ability to recover an amount of waste estimated at 760,000 tons/year (Table 6), in which paper is the essential part in the possibility of recovery and recycling with a quantity of 385,000 tons/year. There are over 2 million tons of plastic packaging products in Algeria by 192 units, but only 4,000 tons are recovered (0.0002%).

TABLE 6: Recycling capacity

Types of waste	Quantity (tons/year)
Paper	385,000
Plastic	130,000
Metals	100,000
Glass	50,000
Various materials	95,000
Total	760,000

5.17 VALORIZATION OF MSW

Valorization is the conversion of waste to energy, fuels, and other useful materials with particular focus on environmental indicators and sustainability goals. It is part of the larger endeavor of loop closing.

5.18 PHYSICOCHEMICAL PROPERTIES OF MSW

Knowledge of the physicochemical parameters of MSW allows the evaluation of the potentially harmful risks of pollution on the environment and human health. Also, it allows the determination of the best ways for the valorization of waste. The most important conditioning parameters for valorization are listed in the following sections [17,18].

Bulk density: An important characteristic of biomass materials is their bulk density or volume. The importance of the bulk density is in relation to transport and storage costs.

Level of moisture: This represents the quantity of water in the MSW.

LCV: This is the total energy content released when the fuel is burnt in air, including the latent heat contained in the water vapor and, therefore, represents the maximum amount of energy potentially recoverable from a given biomass source content, or heat value, released when burnt in air.

Amount of ash: This is the solid residue from the bioconversion in addition to other physicochemical properties such as volatile matter content, C/N ratio, and pH.

From previous studies which were carried out by Tabet [3], Guermoud [19], and Loudjani [9] show the values of physicochemical parameters as shown in Table 7.

TABLE 7: Physicochemical properties of MSW

Property	Value
pH	6 to 7
Medium bulk density	0.45 to 0.55 tons/m3
Fraction of waste >40mm	80% to 90%
Moisture content	50% to 60%
Volatile matter	40% to 60%
Ashes	15% to 40%
C/N	18 to 20
HCV	1,400 to 1,600 kcal/kg

5.19 WASTE-TO-ENERGY CONVERSIONS

Energy from waste is not a new concept, but it is a field which requires a serious attention. There are various energy conversion technologies available to get energy from solid waste, but the selection is based on the physicochemical properties of the waste, the type and quantity of waste feedstock, and the desired form of energy. Conversion of solid waste to energy is undertaken using three main process technologies: thermochemical, biochemical, and mechanical extraction [20].

Biochemical conversion: Biochemical conversion processes make use of the enzymes of bacteria and other microorganisms to breakdown biomass. Biochemical conversion is one of the few which provide environment friendly direction for obtaining energy fuel from MSW. In most of the cases, microorganisms are used to perform the conversion process: anaerobic digestion and fermentation.

1. Anaerobic digestion is the conversion of organic material directly to a gas, termed biogas, which has a calorific value of around 20 to 25 MJ/Nm^3 with methane content varying between 45% and 75% and the remainder of CO_2 (biomass conversion) with small quantities of other gasses such as hydrogen.
2. Fermentation is used commercially on a large scale in various countries to produce ethanol from sugar crops. This produces diluted alcohols which then are needed to be distilled and, thus, suffers from a lower overall process performance and high plant cost.

Thermochemical conversion: Thermal conversion is the component of a number of the integrated waste management solutions proposed in the various strategies. Four main conversion technologies have emerged for treating dry and solid waste: combustion (to immediately release its thermal energy), gasification, pyrolysis, and liquefaction (to produce an intermediate liquid or gaseous energy carrier).

1. Combustion is the burning of biomass in air. It is used over a wide range of commercial and industrial combustion plant outputs to convert the chemical energy stored in the solid waste into heat or electricity using various items of process equipment, such as boilers and turbines. It is possible to burn any type of biomass, but in practice, combustion is feasible only for biomass with a moisture content <50%, unless the biomass is pre-dried.
2. Gasification process means treating a carbon-based material with oxygen or steam to produce a gaseous fuel. Gas produced can be cleaned and burned in a gas engine or transformed chemically into methanol that can be used as a synthetic compound.

3. Pyrolysis is the heating of biomass in the absence of oxygen and results to liquid (termed bio-oil or bio-crude), solid, and gaseous fractions in varying yields depending on a range of parameters such as heating rate, temperature level, particle size, and retention time.
4. Liquefaction is the low-temperature cracking of biomass molecules due to high pressure and results in a liquid-diluted fuel. The advantage of this process, employing only low temperatures of around 200°C to 400°C, has to compete with comparably low yields and extensive equipment prerequisites to provide the pressure levels needed (50 to 200 bars).

Mechanical extraction: It can be used to produce oil from the seeds of solid waste. Rapeseed oil can be processed further by reacting it with alcohol using a process termed esterification to obtain biodiesel, for example.

5.20 BIOGAS MARKET OPTIONS

The kind of energy produced from the biogas depends directly on the needs of the buyer, and there are three different forms: electricity generation, heat and steam generation, and transportation fuel [21].

Electricity generation: This is the most common form of energy produced in facilities constructed today.

1. Combined heat and power (CHP) generation, also known as cogeneration, is an efficient, clean, and reliable approach to generating power and thermal energy from solid waste. By installing a CHP system designed to meet the thermal and electrical base loads of a facility, CHP can greatly increase the facility's operational efficiency and decrease energy costs. At the same time, CHP reduces the emission of greenhouse gasses, which contribute to global climate change.
2. Fuel cell technology: Converting biogas to electricity via fuel cell technology offers significant increases in efficiency and, hence, is a highly desirable technology. Some biogas installations do ex-

ist, utilizing molten carbonate fuel cell technology; however, it is widely considered that solid oxide fuel cell technology is the most promising future technology due to its much higher power density and its applicability to a wide range of scales.

3. Biogas engines: Biogas can be used as a motive power for the production of electricity using engines. A biogas-fueled engine generator will normally convert 18% to 25% of the biogas to electricity, depending on engine design and load factor.
4. Microgas turbines: Small gas turbines that are specifically designed to use biogas are also available. An advantage to this technology is lower NOx emissions and lower maintenance costs; however, energy efficiency is less than with IC engines and it costs more.

Heat and steam generation: Producing and selling heat and steam requires the existence of available industrial customers and matching the supply with their needs. It is also possible to use steam at institutional or domestic complexes.

Transportation fuel: Biogas is used as a transportation fuel in a number of countries. It can be upgraded to natural gas quality in order to be used in normal vehicles designed to use natural gas.

5.21 WASTE-TO-ENERGY TECHNOLOGY IN ALGERIA

In Algeria, a little interest was given to this technology despite the important resources and the successful applications that have been made by the National Institute of Agronomy (El Harrach) and the CDER through the establishment of two experimental plants in Bechar and Ben Aknoun for the study of biogas production from cow dung. However, we must note the efforts of the Algerian government in these last years to develop this technology by upgrading the landfill of Ouled Fayet which has been put into operation in 2011. The project's main objective is the capture of the landfill's gas which contains 50% of methane (CH_4); the expected amount of emission reduction is 83,000 T equivalent CO_2/year [22].

In addition to biomass, power projects are at the feasibility study stage such as the Sonelgaz's biomass power project in the Oued Smar site, which has an installed capacity of 2 MW that can reach a peak of 6 MW from the discharge of this site, and the energy recovery plant of biogas generated in the landfill of Batna [12].

5.22 WASTE-TO-ENERGY-RELATED ENVIRONMENTAL ISSUES

5.22.1 REDUCTION IN LANDFILL DUMPING

Landfills require large amounts of land that could be used for other purposes; incineration of solid waste can generate energy while reducing the volume of waste by up to 90%.

5.22.2 REDUCED DEPENDENCE ON FOSSIL FUELS

With advanced technologies, waste can be used to generate fuel that does not require mining or drilling for increasingly scarce and expensive non-renewable fossil-fuel resources.

5.22.3 REDUCED GREENHOUSE-GAS EMISSIONS AND POLLUTION

Using waste as a feedstock for energy production reduces the pollution caused by burning fossil fuels. While traditional incineration still produces CO_2 and pollutants, advanced methods such as gasification, pyrolysis, and liquefaction, have the potential to provide a double benefit: reduced CO_2 emissions compared with incineration or coal plants, and reduced methane emissions from landfills.

5.22.4 WTE PROVIDES CLEAN ENERGY

The WTE technology has significantly advanced with the implementation of the Clean Air Act, dramatically reducing all emissions.

5.23 WASTE-TO-ENERGY CHALLENGES

5.23.1 LACK OF VERSATILITY

Many waste-to-energy technologies are designed to handle only one or a few types of waste (biomass, solid waste or others). However, it is often impossible to fully separate different types of waste or to determine the exact composition of a waste source. For many, waste-to-energy technologies to be successful, they will also have to become more versatile or be supplemented by material handling and sorting systems.

5.23.2 WASTE-GAS CLEANUP

The gas generated by processes like pyrolysis and thermal gasification must be cleaned of tars and particulates in order to produce clean and efficient fuel gas.

5.23.3 CONVERSION EFFICIENCY

Some waste-to-energy pilot plants, particularly those using energy-intensive techniques like plasma, have functioned with low efficiency or actually consumed more energy than they were able to produce.

Toxic materials include trace metals such as lead, cadmium and mercury, and trace organics, such as dioxins and furans. Such toxins pose an environmental problem if they are released into the air with plant emissions or if they are dispersed in the soil, allowed to migrate into ground water supplies, and work their way into the food chain. The control of such toxins and air pollution is the key feature of environmental regulations governing MSW-fueled electric generation.

5.23.4 REGULATORY HURDLES

The regulatory climate for waste-to-energy technologies can be extremely complex. At one end, regulations may prohibit a particular method, typically incineration, due to air-quality concerns, or classify ash byproducts of waste-to-energy technologies as hazardous materials. At the other end, while changes in the power industry have allowed small producers to compete with established power utilities in many areas, the electrical grid is still protected by yet more regulations, presenting obstacles to would-be waste-energy producers.

5.23.5 HIGH CAPITAL COSTS

Waste-to-energy systems are often quite expensive to install. Despite the financial benefits they promise due to reductions in waste and production of energy, assembling the financing packages for installations is a major hurdle, particularly for new technologies that are not widely established in the market.

5.24 CONCLUSION

This paper gives an overview on the Algerian potential of solid waste including MSW, ISW, and HW as biomass sources. The management of solid waste (MSW) and valorization is based on the understanding of MSW composition by its categories and physicochemical characteristics.

Energy from waste is not a new concept, but it is a field which requires a serious attention. There are various energy conversion technologies (thermochemical, biochemical, and mechanical extraction) to produce useful products (electricity, heat, and transportation fuel).

In general, the government should, first and foremost, implement its own decisions and work towards encouraging independent renewable energy producers, in general, and energy generation by WTE technologies, in particular. By doing so, the overall energy generation capacity will increase, the dependence of Algeria on imported fossil fuels will be reduced, and a significant reduction in pollution and greenhouse gas emissions will occur.

5.25 RECOMMENDATIONS

The recommendations of this research are the following:

- Solid waste can be used as an energy source in Algeria. However, the WTE facilities must operate under strict standards, which will minimize environmental impact and adhere to the precaution principle.
- Implementation of landfill disposal techniques should be encouraged for the valorization of biogas.
- Waste-to-energy and valorization of Algerian solid waste is a new subject that needs to be developed.

REFERENCES

1. Sonelgaz Group Company
2. Stambouli, AB: Algerian renewable energy assessment: the challenge of sustainability. Energy Policy. 39, 4507–4519 (2011).
3. Stambouli, AB: Promotion of renewable energies in Algeria: strategies and perspectives. Ren. Sust. Energy Reviews. 15, 1169–1181 (2011).
4. Population Reference Bureau (PRB): World population data sheet (2011) http://www.prb.org webcite (2011). Accessed 2011
5. BP Statistical Review of World Energy: Statistical review, London
6. Boudries, R, Dizene, R: Potentialities of hydrogen production in Algeria. Int. J. Hydrogen Energy. 33, 4476–4487 (2008).
7. Himri, Y, Arif, SM, Boudghene, SA, Himri, S, Draoui, B: Review and use of the Algerian renewable energy for sustainable development. Ren. Sust. Energy Reviews. 13, 1584–1591 (2009).
8. Djemaci, B, Chertouk, MAZ: La gestion intégrée des déchets solides en Algérie. Contraintes et limites de samise en œuvre. International Centre of Research and Information on the Public
9. Loudjani, F: Guide des techniciens communaux pour la gestion des déchets ménagers et assimiles. The Ministry of Land Planning and the Environment (MATE)

10. Gourine, L: Country report on the solid waste management: Algeria. The regional solid waste exchange of information and expertise network in Mashreq and Maghreb countries
11. Ouzir, M: Gestion ecologique des déchets solides industriels: casd'étude la villed'arzew, University of M'sila (2008)
12. Louai, N: Evaluation énergétique des déchets solides en Algérie, une solution climatique et un nouveau vecteur énergétique, University of Batna (2009).
13. Bendjoudi, Z, Taleb, F, Abdelmalek, F, Addou, A: Healthcare waste management in Algeria and Mostaganem department. Waste Manage. 29, 1383–1387 (2009).
14. Sefouhi, L, Kalla, M, Aouragh, L: Health care waste management in the hospital of Batna City (Algeria), Paper presented at the Singapore International Conference on Environment and BioScience, Singapore (2011)
15. Redjal, O: Vers un développement urbain durable: Phénomène de prolifération des déchets urbains et stratégie de préservation de l'écosystème: exemple de Constantine, University of Mentouri Constantine (2005).
16. Kehila, Y, Mezouari, F, Matejka, G: Impact de l'enfouissement des déchets solides urbains en Algérie: expertise de deux Centresd' Enfouissement Technique (CET) à Alger et Biskra. Revue Francophone d'Ecologie Industrielle Déchets, Sciences & Techniques. 56, 29–38 (2009)
17. Alamgir, M, Ahsan, A: Characterization of MSW and nutrient contents of organic component in Bangladesh. EJEAFChe. 6(4), 1945–1956 (2007)
18. McKendry, P: Energy production from biomass (part 1): overview of biomass. Bioresour Technol. 83, 37–46 (2002).
19. Guermoud, N, Ouadjnia, F, Abdelmalek, F, Taleb, F, Addou, A: Municipal solid waste in Mostaganem City (Western Algeria). Waste Manage. 29, 896–902 (2009).
20. McKendry, P: Energy production from biomass (part 2): conversion technologies. Bioresour Technol. 83, 47–54 (2002).
21. Münster, M, Lund, H: Use of waste for heat, electricity and transport–Challenges when performing energy system analysis. Energy. 34, 636–644 (2009).
22. Tabet, MA: Types de traitement des déchets solides urbains: evaluation des coûts et impacts sur l'environnement. Rev. Energ. Ren. Special number Production et Valorisation –Biomasse. 2001, 97–102 (2001)

CHAPTER 6

Energy Recovery Routes from Municipal Solid Waste: A Case Study of Arusha-Tanzania

ARTHUR OMARI, MAHIR SAID, KAROLI NJAU, GEOFFREY JOHN, AND PETER MTUI

6.1 INTRODUCTION

Municipal solid waste generation has been in the increase due to population growth, changing lifestyles, technology development and increased consumption of goods. The increase of wastes generation may lead to environmental problems if not properly managed (A. Johari et al., 2012). Urban centers in developing countries are facing a challenge in solid waste management due to population growth and are constrained by lack of an effective recycling of the biodegradable components into useful materials, poor waste management and waste handling infrastructure (R.K. Henry et al., 2006; J.-H. Kuo et al., 2008).

Despite having abundant solid waste in developing countries, these countries are facing energy crisis which pose a challenge to their economic and social development. Combining waste management with waste energy recovery step from municipal solid waste can address the problems of solid waste management and partly the energy crisis. A disposal method using thermal degradation processing could be a better option for the waste management than biogenic methods. This method has advantages, such as substantial reduction in volume and mass. In order to apply the method in a large scale, there are fundamental parameters such as fuel behavior in thermal degradation, energy contents and its chemical reactions that should be in place so as to assist designers to come up with an appropriate method of waste energy recovery and disposal system (M.J. Quina et al., 2008).

In this study, the thermal degradation behavior of municipal solid waste in a growing urban city of Arusha, Tanzania as a case study is undertaken. This includes determination of its proximate analysis, ultimate analysis higher heating value and kinetics.

6.2 MATERIAL AND METHODS

6.2.1 METHODOLOGY

The methodology consists of sampling selection, sorting and laboratory analysis to determine the chemical and physical properties of municipal solid waste of Arusha city. The method of sampling was based on ASTM D5231 namely random truck sampling and quatering (A. AbdAlqader and J. Hamad, 2012). In this study the wastes were collected by means of push carts and donkey carts, and were randomly collected from different collecting point of Sakina, Kaloleni and Central market within the Arusha City. The wastes were sorted and weighted by using weighing balance and then separated according to defined classification such as plastics, glass, paper, food waste and metals. The non-combustible wastes were removed from the rest of the wastes. The combustible waste was availed for analysis in accordance to the method developed by (P. McCauley-Bell et al., 1997; G.S. Yang, 2012).

In order accurately to get waste composition an average weight of about 200kg of municipal solid waste was taken. The waste was then taken as good representative of the total municipal solid waste composition at each collecting points under this study. The samples were subjected to standards test methods of proximate and ultimate analysis in accordance to ASTMD3172 and ASTM D3176 respectively. (ASTM;ASTM)

The thermal degradation analysis was studied under Nitrogen condition using a thermo gravimetric analyzer type NETZSCH STA 409 PC Luxx connected to power unit 230V, 16A. High purity nitrogen, 99.95% used as carrier gas controlled by gas flow meter was fed into the thermo gravimetric analyzer with flow rate of 60ml/min and a pressure of 0.5 bars. In the STA 409 PC Luxx, proteus software was utilized to acquire, store and analyze the data.

6.2.2 SAMPLE PREPARATION

The samples were shredded into smaller pieces of approximately 30mm size, mixed and grounded in a grinding machine to less than 1mm size, this is in order to increases surface area of the sample that will allow easier penetration of heat (M.H.M. Yusoff and R. Zakaria, 2012) Then a sample of 30±0.1 mg with average particle size less than 1mm was loaded to crucible and subjected into furnace and heated from 303 to 1273 K at heating rate of 10 K/min, 20 K/min, 30 K/min and 40 K/min. The heating rate variation changes the peak temperature of the decomposition, as the heating rate increases, the peak temperature also increases (S. Ledakowicz and P. Stolarek, 2003). The calculated thermo-gravimetric output from proteus software was obtained as thermal decomposition profile; thermo-gravimetric (TG), differential thermo-gravimetric (DTG) and differential scanning calorimetry (DSC) curves.

Heat release and absorbed by municipal solid waste degradation was determined by using differential scanning calorimetry curves. The DSC monitors heat effect associated with phase changes transitions and chemical reactions as a function of temperature (R. Huffman and W.-P. Pan, 1990). The heat was determined by calculating the area between the baseline and the curve. The heat can be positive or negative. When the heat is positive the process is endothermic and when the heat is negative the process is exothermic (M. Tettamanti et al., 1998)

6.2.3 THE KINETIC PARAMETER.

The kinetic parameter was determined by using Kissinger's method. This is used as a standard method for studying the thermal degradation of municipal solid waste under non isothermal condition (S. Ledakowicz and P. Stolarek, 2003). The rate constant is expressed by Arrhenius Equation (1) where, k is the rate constant, which is temperature dependent (T. Sonobe and N. Worasuwannarak, 2008).

$$k = A \exp(-E_a/RT) \tag{1}$$

$$dx/dt = Af(x) \exp(-E_a/RT) \tag{2}$$

$$x = (w_0 - w_n) / (w_0 - w_\infty) \tag{3}$$

where, x is the reacted fraction, w_0 the initial mass, w_t the mass remaining at time t, w_∞ the final mass , T the absolute temperature, E_a the activation energy, A the pre-exponential factor, R the universal gas constant and f(x) the algebraic function depending on the reaction mechanism. The temperature rise at a constant heating rate (β) is expressed as shown in Equation 4.

$$\beta = dT/dt \tag{4}$$

Equation 5 is a result of differentiation of Equation 2

$$d^2x/dt = \{E_a\beta/RT^2 + Af'(x)\exp(-E_a/RT)\}\ dx/dt \tag{5}$$

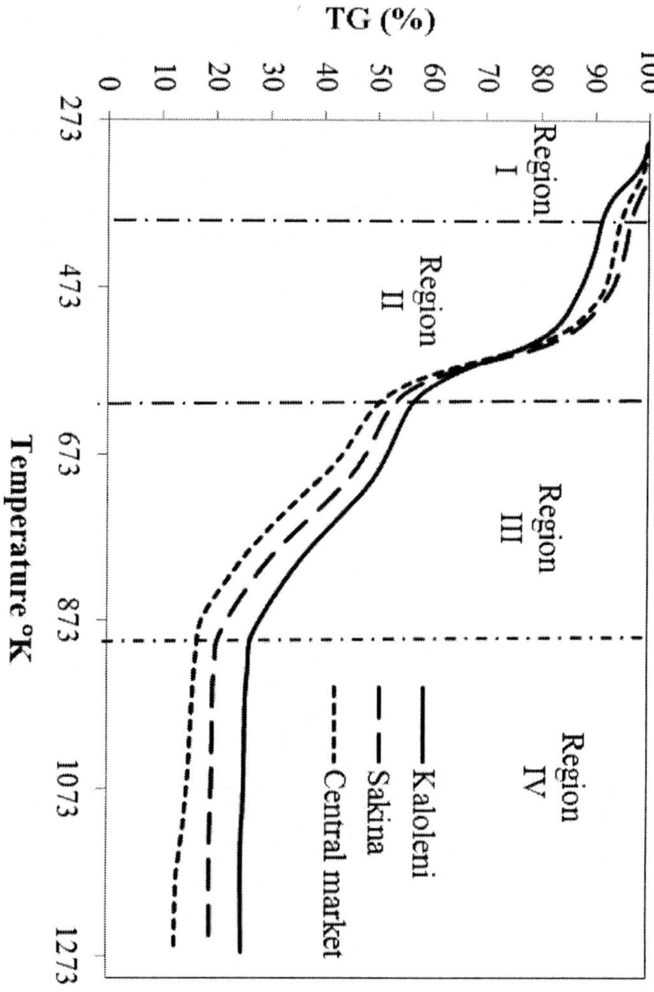

FIGURE 1: TG of Municipal solid waste

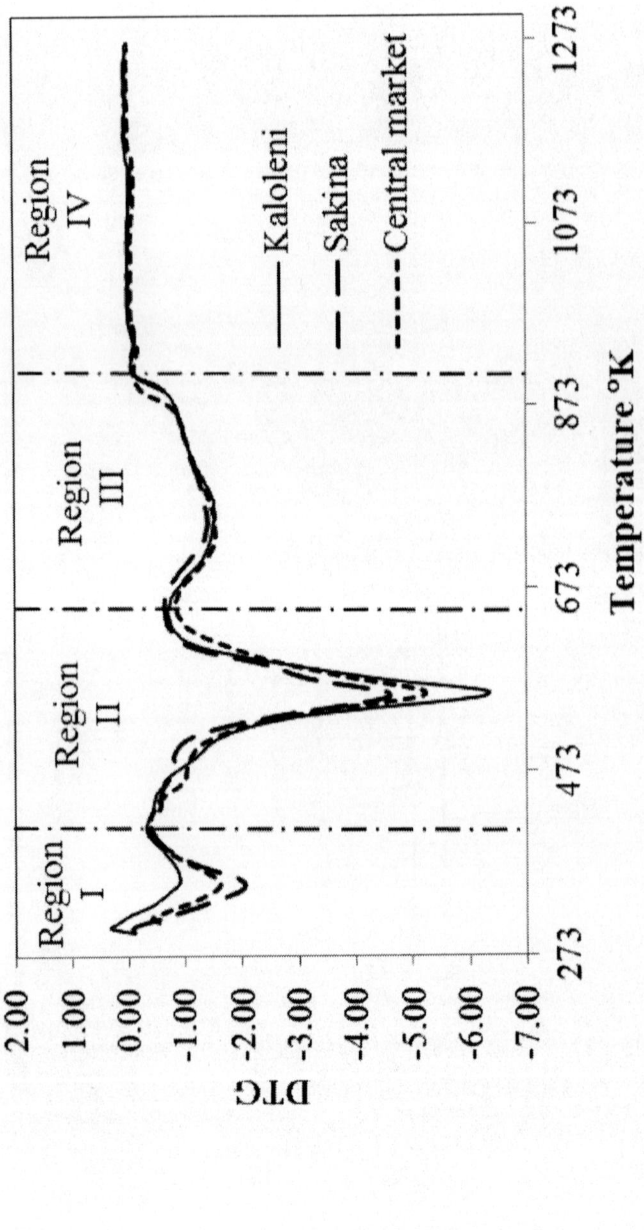

FIGURE 2: DTG of municipal solid waste

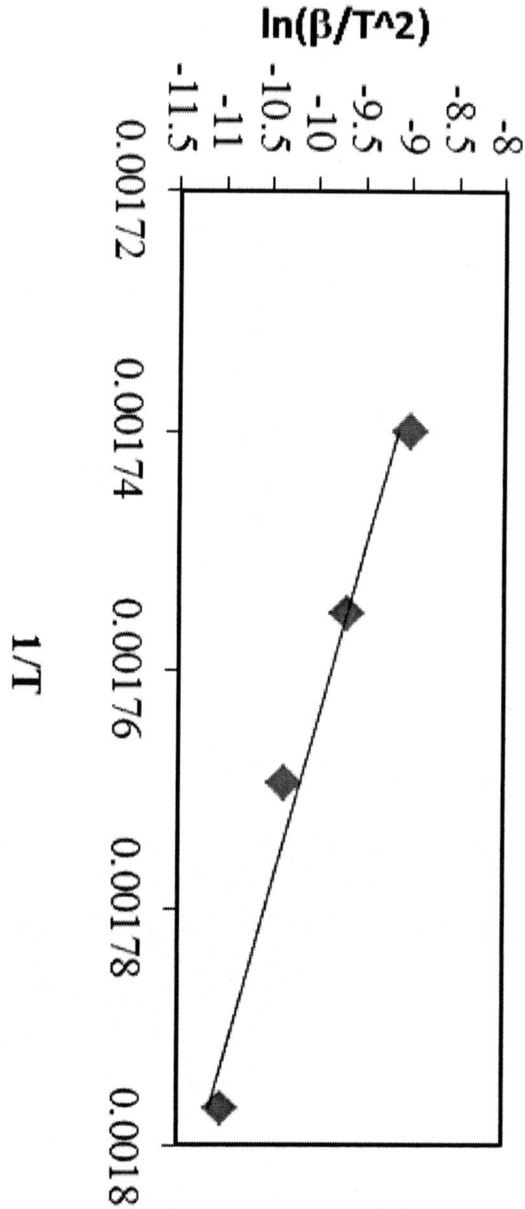

FIGURE 3: Determination of kinetic parameter of Arusha municipal solid waste.

The maximum rate occurs at a temperature T_{peak}; approximations at T_{peak} condition yield Equation 6.

$$\ln(\beta/T_{peak}^{2}) = \ln(AR/E_a) - (E_a/RT_{peak}) \quad (6)$$

Equation 6 is a straight line graph, of $\ln(\beta/T_{peak}^{2}$ v/s $(1/T_{peak})$, The line slope is E_a/R and the intercept on the vertical axis is $\ln(AR/E_a)$, which are used to determine the values of E_a and A.

The fractional pyrolysis of municipal solid waste component is obtained by taking the ratio of the change mass of municipal solid waste component at time t and total reactive mass of a sample as shown in Equation 3.

6.3 RESULTS AND DISCUSSION

6.3.1 PROXIMATE AND ULTIMATE ANALYSIS

The results of proximate and ultimate analysis are shown in Table 1. The moisture content of the municipal solid waste as received ranges between 55.70 and 63.99 wt. %, which is more than 50 wt. % of the total weight of the sample. This high moisture content is prohibitive for combustion process as it rises the ignition temperature, also its contents reduces the calorific value of the fuel (M. Muthuraman et al., 2010), the moisture could be reduced by drying. The volatiles released on dry basis of MSW for Kaloleni, Sakina and Central market are 74.43, 84.00 and 78.31 wt.%, respectively, whilst the volatile matter contained in pure biomass such as forest residue, oak wood, and pine are 79.9, 78.1 and 83.1 wt. % respectively (S.V. Vassilev et al., 2010). Generally, fuels that contains high volatile, have low fixed carbon, the case is same for the municipal solid waste from Kaloleni which has fixed carbon of about 17 wt. %, which is higher than that of Sakina and Central market. The advantage of high volatile and low fixed carbon is rapid burning of a fuel, while a fuel with low volatile and high fixed carbon like coal need to be burn on a grate as it take long time to burn out, unless it is pulverized to a very small size (P. McKendry, 2002) Therefore the value of volatile matter and fixed carbon shows that the municipal solid waste is combustible. The ash range between 3.29 to 5.97 wt. %, which is small, this is advantage to waste management and environment because the possibility of having small quantity of heavy metals, salts, chlorine and organic pollutant is small (C.H. Lam et al., 2010). The ultimate analysis of the municipal solid waste shows that the concentration of phosphorus and chlorine are negligible, the carbon and hydrogen content were above 50% and 5% respectively. The oxygen content was more than 34%. Sulfur is about 0.29%, this is low compared to values from 1.1 wt. % of bituminous coal analysis (T. Nakao et al., 2006).

TABLE 1: Proximate, ultimate analysis and HHV of Arusha municipal solid waste

	Proximate analysis						
Location	Moisture of received MSW (wt. %)	Volatile (wt.%) dry basis	Ash (wt. %) dry basis	Fixed carbon (wt. %) dry basis	HHV (MJ/kg)		
Kaloleni	59.67	74.43	8.16	17.41	11.90		
Sakina	63.99	84.00	10.00	6.00	11.37		
Central market	55.70	78.30	13.48	8.22	12.76		
	Ultimate analysis						
Location	C (wt. %)	H (wt. %)	O (wt. %.)	N (wt. %)	S (wt. %)	Cl (wt. %)	P (wt. %)
Kaloleni	55.57	5.34	34.88	2.09	0.31	0.04	0.10
Sakina	55.70	5.29	34.27	2.13	0.22	0.07	0.13
Central Market	53.20	5.24	34.71	2.86	0.37	0.04	0.11

6.3.2 CALORIFIC VALUE

The municipal solid waste calorific value is about 12 MJ/kg. This value is smaller than average biomass heating value of about 17MJ/kg (F. Heylighen, 2001) This means energy release during combustion of MSW is smaller compared to biomass combustion. This means that one needs to burn larger amount of MSW to get the same amount of energy. The energy content of MSW can be improved by pre-treating the MSW so as to reduce oxygen amount, since oxygen reduces the energy content of a fuel (P. McKendry, 2002). The MSW can be cofired with coal for improving energy content (M. Sami et al., 2001;Z. Li et al., 2004). Other processes to improve energy content of MSW are pyrolysis, gasification or torrefaction, these are used to produce bio-oil, syngas or char respectively.

The municipal solid waste from all collecting points degraded to 75 to 85 wt. % in the thermo gravimetric analyser as shown in Figure 1. The MSW from Central market degraded by 85 wt. %, while the Kaloleni degraded by 75 wt. %. The residue formed is between 25 and 15 wt. %. The residue contains fixed carbon and ash, the high residue is observed at MSW from Kaloleni (25 wt. %) and the lowest residue is observed at MSW from Central market and Sakina 15 wt. %. The char can be used as a fuel, but MSWs that have high ash content hinder the combustion of char due to the layer of ash formed on the surface it inhibited the diffusion of oxygen into the char (D.A. Himawanto et al., 2013).

6.3.3 DTG CURVES

Figure 2 shows the derivative of thermo-gravimetric analysis (DTG), which has four visible regions; these are moisture release region, lignocellulosic degradation region, plastic degradation region and char pyrolysis region (Z. Lai et al., 2011).

The moisture release region is ranging between 303 and 423. Lignocellulosic degradation region ranges between 423 and 643 K, at these region volatile matters are released; the region corresponds to pyrolysis of lignocellulosic biomass. The plastic degradation ranges between 643 and 913 K and the char pyrolysis region ranges between 913 and 1273. The same identified regions were also observed by Lai et al., (2011).

DTG curves at different heating rate were used to develop Figure 3, which was used to calculate the activation energy (Ea) and pre exponential factor (A), as given in Table 2. The activation energy of MSW ranged between 205.934 kJ/mol and 260.60kJ/mol. This value is higher than that of biomass and coal which range between 50 and 180kJ/mol and 30 and 90 kJ/mol respectively. This corresponds to the biomass of cypress wood chips and macadamia nut shells as observed by Vhathvarothai et al. (2013), that the value was 168.7kJ/mol and 164.5kJ/mol respectively (N. Vhathvarothai et al., 2013). This shows that MSW need high energy to react as compared to biomass and coal. The reactivity of MSW can be increased by reducing the noncombustible material such as oxygen and also to remove volatile material, these can be done by pretreating the material through torrefaction process (A.K. Biswas, 2011).

TABLE 2: Activation energy and Pre-exponential factor of municipal solid waste

Location	E_a (kJ/mol)	A (s^{-1})
Kaloleni	258.680	9.142 x 1023
Sakina	205.934	8.977 x 1018
Central Market	260.60	1.186 x 1028

6.3.4 DSC CURVES

The differential scanning calorimetry (DSC) curves shown in Figure 4, reveal endothermicity between 303 and 423, this is due to evaporation of moisture. In the temperature range of 423 to 1273 K the process undergoes exothermic reaction due to the devolatilization of the municipal solid waste and plastic pyrolysis. The energy absorbed due to evaporation of moisture for wastes from Kaloleni, Sakina and Central market collecting points were 0.11 MJ/kg, 0.20 MJ/kg and 0.15 MJ/kg respectively, whilst energy released from the same respective collection points were -7.6MJ/kg, -8.3 MJ/kg and-8.5 MJ/kg in respective manner. The energy released in the DSC by municipal solid waste was lower than higher heating value (12.54 MJ/kg). This is because the DSC used nitrogen as heating media while in bomb calorimeter oxygen is applied for combustion.

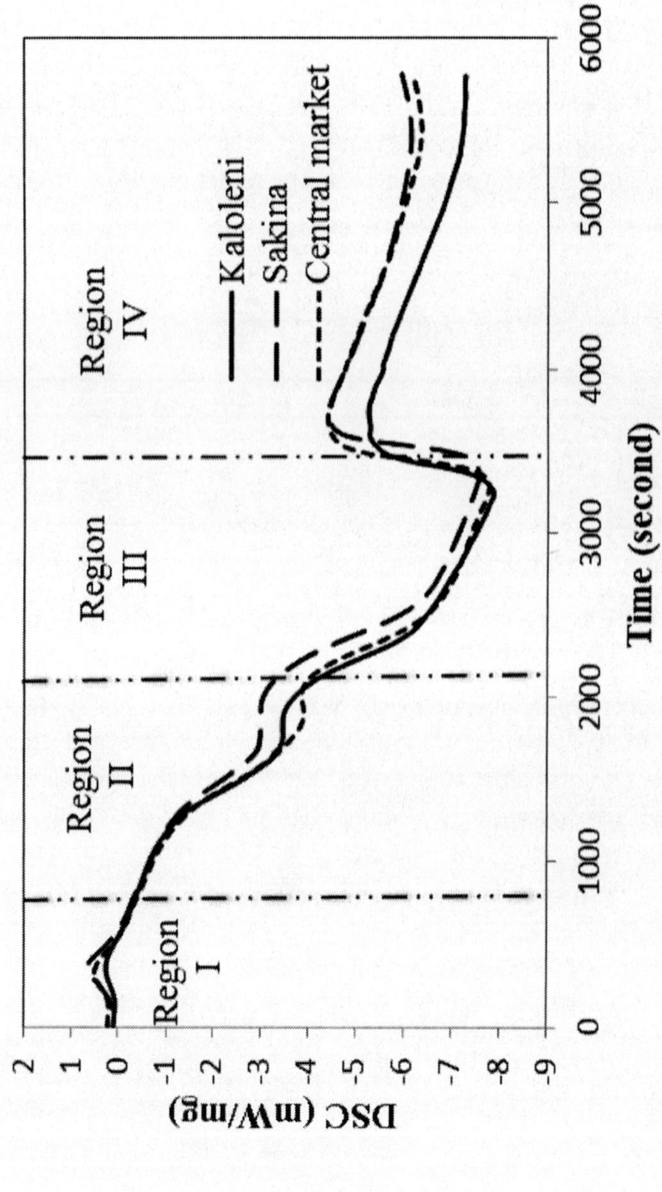

FIGURE 4: DSC of Arusha Municipal solid waste sites.

6.4 CONCLUSION

This paper presents finding related to municipal solid waste characterization of Arusha city. The proximate analysis of municipal solid waste show that, the waste contains more than 50% and 5% of carbon and hydrogen respectively which may contribute to high calorific value of Arusha municipal solid waste. The ultimate analysis shows that average amount of nitrogen, sulfur, chlorine and phosphorus are small, these reduce emissions during combustion.

The energy content of waste determined by bomb calorimeter is about 12MJ/kg this is about 30% of energy containing in coal and 60% of energy containing in biomass. The activation energy was ranging between 205.9 and 260.6kJ/mol. The municipal solid waste shows exothemicity property at the devolatilization zone. The devolatization zone shows that the municipal solid waste can be easily ignited at temperature above 423 K. Therefore municipal solid waste has a good potential to be used as a fuel.

REFERENCES

1. AbdAlqader, A. and Hamad, J. (2012). Municipal solid waste composition determination supporting the integrated solid waste management in Gaza strip. Int. J. Environ. Sci. Dev. 3(2): 172-177.
2. ASTM D3172-07 in Standard Practice for Proximate Analysis of Coal and Coke. West Conshohocken, PA: 19428-2959.
3. ASTM D3176-89 in Standard Test Method for Ultimate Analysis of Coal and Coke. ASTM International, West Conshohocken, PA.
4. Biswas, A.K. (2011). Thermochemical behavior of pretreated biomass.
5. Henry, R.K., Yongsheng, Z. and Jun, D. (2006). Municipal solid waste management challenges in developing countries – Kenyan case study. Waste Management. 26(1): 92-100.
6. Heylighen, F. (2001). Encyclopedia of Life Support Systems.
7. Himawanto, D.A., Saptoadi, H. and Rohmat, T.A. (2013). Thermogravimetric Analysis of Single-Particle RDF Combustion. Modern Applied Science. 7(11).
8. Huffman, R. and Pan, W.-P. (1990). Combining DSC and TG data for measuring heats of reaction. Thermochimica acta. 166(251-265.
9. Johari, A., Hashim, H., Mat, R., Alias, H., Hassim, M. and Rozzainee, M. (2012). Generalization, formulation and heat contents of simulated MSW with high moisture content. Journal of Engineering Science and Technology. 7(6): 701-710.

10. Kuo, J.-H., Tseng, H.-H., Rao, P.S. and Wey, M.-Y. (2008). The prospect and development of incinerators for municipal solid waste treatment and characteristics of their pollutants in Taiwan. Applied Thermal Engineering. 28(17): 2305-2314.
11. Lai, Z., Ma, X., Tang, Y. and Lin, H. (2011). A study on municipal solid waste (MSW) combustion in N2/O2 and CO2/O2 atmosphere from the perspective of TGA. Energy. 36(2): 819-824.
12. Lam, C.H., Ip, A.W., Barford, J.P. and McKay, G. (2010). Use of incineration MSW ash: A review. Sustainability. 2(7): 1943-1968.
13. Ledakowicz, S. and Stolarek, P. (2003). Kinetics of biomass thermal decomposition. Chemical papers - slovak Academic of Science. 56(6): 378-381.
14. Li, Z., Lu, Q. and Na, Y. (2004). N2 O and NO emissions from co-firing MSW with coals in pilot scale CFBC. Fuel Processing Technology. 85(14): 1539-1549.
15. McCauley-Bell, P., Reinhart, D.R., Sfeir, H. and Ryan, B.O.T. (1997). Municipal solid waste composition studies. Practice Periodical of Hazardous, Toxic, and Radioactive Waste Management. 1(4): 158-163.
16. McKendry, P. (2002). Energy production from biomass (part 1): overview of biomass. Bioresource Technology. 83(1): 37-46.
17. Muthuraman, M., Namioka, T. and Yoshikawa, K. (2010). A comparative study on co-combustion performance of municipal solid waste and Indonesian coal with high ash Indian coal: A thermogravimetric analysis. Fuel Processing Technology. 91(5): 550-558.
18. Nakao, T., Aozasa, O., Ohta, S. and Miyata, H. (2006). Formation of toxic chemicals including dioxin-related compounds by combustion from a small home waste incinerator. Chemosphere. 62(3): 459-468.
19. Quina, M.J., Santos, R.C., Bordado, J.C. and Quinta-Ferreira, R.M. (2008). Characterization of air pollution control residues produced in a municipal solid waste incinerator in Portugal. Journal of Hazardous Materials. 152(2): 853-869.
20. Sami, M., Annamalai, K. and Wooldridge, M. (2001). Co-firing of coal and biomass fuel blends. Progress in Energy and Combustion Science. 27(2): 171-214.
21. Sonobe, T. and Worasuwannarak, N. (2008). Kinetic analyses of biomass pyrolysis using the distributed activation energy model. Fuel. 87(3): 414-421.
22. Tettamanti, M., Collina, E., Lasagni, M., Pitea, D., Grasso, D. and La Rosa, C. (1998). Characterization of fly ash from municipal solid waste incinerators using differential scanning calorimetry. Thermochimica acta. 321(1): 133-141.
23. Vassilev, S.V., Baxter, D., Andersen, L.K. and Vassileva, C.G. (2010). An overview of the chemical composition of biomass. Fuel. 89(5): 913-933.
24. Vhathvarothai, N., Ness, J. and Yu, Q.J. (2013). An investigation of thermal behaviour of biomass and coal during copyrolysis using thermogravimetric analysis. International Journal of Energy Research.
25. Yang, G.S. (2012). Identification of the Municipal Solid Waste Characteristics and Potential of Plastic Recovery at Bakri Landfill, Muar, Malaysia. Journal of Sustainable Development. 5(7).
26. Yusoff, M.H.M. and Zakaria, R. (2012). Combustion of Municipal Solid Waste in Fixed Bed Combustor for Energy Recovery. Journal of Applied Sciences. 12(11).

PART IV

RECYCLING AND REUSE

CHAPTER 7

Recycling of Pre-Washed Municipal Solid Waste Incinerator Fly Ash in the Manufacturing of Low Temperature Setting Geopolymer Materials

CLAUDIO FERONE , FRANCESCO COLANGELO,
FRANCESCO MESSINA, LUCIANO SANTORO,
AND RAFFAELE CIOFFI

7.1 INTRODUCTION

The incineration of municipal solid wastes has relevant social, economic and environmental impacts. This process generates several gaseous effluents and produces both solid and liquid residues. The latter corresponds to 10% of initial waste volume. Hence, the proper management of these residues, particularly bottom and fly ash, is of crucial importance, and techniques for their stabilization/solidification need further optimization [1–3].

The aforementioned issue can find an answer in eco-design approaches which aim for material recovery to reduce the consumption of natural raw materials in the field of cement-based materials manufacturing. The unreactive stabilized waste can be employed together with solid wastes produced by other industrial processes. In fact, resource optimization implies significant advantages in terms of economic, energetic and environmental parameters of concrete industry (e.g., LEED—Leadership in Energy and Environmental Design Indicators). This industry is able to recycle and stabilize many kinds of solid wastes both in binder and artificial aggregates production, in order to achieve sustainability objectives [4–12].

Fly ash from municipal solid waste incinerators (MSWI-FA) are classified as hazardous in the European Union. Therefore, prior to a proper stabilization process, their contaminant release has to be evaluated. From this point of view, in addition to heavy metals, chlorides and sulfates pose major issues. In fact, untreated MSWI-FA release very high amounts of these pollutants when they are submitted to the leaching test UNI 10802-2004 [13] that derives from test EN 12457-2: 2002 [14]. These release values are always higher than the limits required for both hazardous and non-hazardous waste landfilling. Furthermore, any effective stabilization process is not economically sound if MSWI-FA are stabilized without a proper washing pretreatment. In this regard, attempts to optimize the solid/liquid ratio, with consequent water consumption reduction, can be found in the literature [15–18]. More specifically, Colangelo et al. [17] have applied one-step and two-step washing pretreatments on three different fly ash samples proving that the use of a very limited 2:1 overall liquid to solid ratio is possible. Thereby, the pre-washed MSWI-FA have been proposed as cement bound granular material in the manufacture of sub-base layer for road construction. Furthermore, a cost analysis of the complete process has been made too. Specifically, this cost analysis was carried out taking into account the charges for cement-stabilization, washing pretreatment and washing salt disposal less the benefits from material reuse and comparing this with the charge for untreated MSWI-FA simple disposal. The results have demonstrated that the proposed process is economically sound.

As far as the stabilizing matrices are considered, it is well known that cementitious ones based on cements, pozzolans, blast furnace slag and lime are often not suitable to reduce the very high mobility of chlorides and sulfates down to the imposed regulation limits. The reason for this is that high chloride and sulfate concentration has a strong negative effect on their efficiency [19,20]. Alternative matrices, such as those based on alkali-activated aluminosilicate binders, including the geopolymers, are worthy of consideration because excellent mechanical properties, durability, resistance to acid attack and thermal stability can be achieved. The synthesis of geopolymers takes place by polycondensation and can start from silicoaluminate and aluminosilicate materials. When they are in contact with the high pH of alkaline solution, raw materials dissolve and the inorganic polymers precipitate [21–25]. Recently, the applications of this broad class of materials in several fields of engineering have been deeply discussed by many authors, revealing a great number of possible technological solutions [26–33]. Great interest also derives from the possibility of employing naturally occurring silicoaluminate and aluminosilicate industrial wastes, such as coal fly ash, blast furnace slag, clay sediment, etc., thus decreasing the environmental impact for the manufacturing of new materials based on geopolymers. In addition, the synthesis of neo-formed phases takes place at low temperatures, not higher than 60 °C [21–25,34–37]. All these considerations imply, in comparison to traditional cement-based materials, a reduction of natural raw materials consumption and greenhouse gases emission, particularly CO_2.

In the field of hazardous solid waste treatment, the above polycondensation phases can favour the entrapment of contaminants, by means of both physical and chemical mechanisms, when geopolymers are employed as stabilizing matrices. Particularly, the stabilization/solidification of MSWI-FA in geopolymers has been already discussed by several authors in recent years [38–43]. Even if geopolymeric matrices setting and hardening are based on a different chemistry, as for the cementitious systems, the negative effect of the presence of chlorides and sulfates on the polycondensation kinetic was observed [44,45].

In order to optimize the entire cycle of MSWI-FA stabilization, water pre-washing can be applied for chlorides (and other soluble salts such as sulfates) removal. To this regards, Zheng et al. [46] investigated the effect of water-wash on geopolymerization. They concluded that a combined washing-stabilization process gave better immobilization efficiency of some heavy metals and higher early strength of hardened specimens. The drawbacks of this approach are the related water consumption for complete chlorides removal and the secondary pollution arising from the transfer of chlorides and other soluble salts from the ash to the washing water. So, the washing pretreatment must be optimized in relation to the minimum washing water requirement and maximum allowed residual amount of chlorides (and other soluble salts). Finally, an adequate binder to waste ratio is to be used in the stabilization process for the economically sound management of MSWI-FA landfilling/reuse.

In this work, coal fly ash has been used for the synthesis of geopolymeric matrices that can incorporate and stabilize three samples of fly ash from municipal solid wastes incinerators (MSWI). The different MSWI-FA samples have been used not only as received, but also after washing to reduce their chloride content. The products obtained under the different experimental conditions have been characterized from the qualitative point of view by means of Fourier transform infrared spectroscopy (FT-IR), X-ray diffraction (XRD) and scanning electron microscopy (SEM) and from the quantitative point of view through the measurement of the amounts of silicate and water reacted upon polycondensation. Finally, density and compressive strength of hardened specimens have also been evaluated and the environmental and technological classification of the final materials has been assessed by means of leaching tests and management considerations. The main goal of this work is gathering experimental data useful for the MSWI-FA management in relation to final reuse/utilization options. From this point of view, it is clear that, due to the intrinsic characteristics of the materials employed, hi-tech solutions will be precluded and options such as abandoned quarry filling or low temperature setting, soft brick manufacturing will be more appropriate.

7.2 MATERIALS AND METHODS

The three MSWI-FA samples come from plants located in southern, central and northern Italy and have been named A, B and C, respectively. These samples have been collected downstream of the air pollution control (APC) device and comprised MSWI fly ash plus APC residue. The three samples have been submitted to chemical analysis, as described extensively in a previous work [17]. These three ash samples have been submitted to total acid digestion according to ASTM 5258-92 [47] and subsequent chemical analysis through inductively coupled plasma atomic emission spectrometry (ICP-AES) technique for the determination of metal contents. Chloride and sulfate content has been determined by means of the Mohr method and ionic liquid chromatography, respectively. All the measurements have been replicated nine times, and when reporting the data, mean values and standard deviations have been shown.

The three MSWI-FA have been characterized in terms of heavy metal, chloride and sulfate release by means of UNI 10802 [13] leaching tests. This is a test that makes use of deionized water with a liquid to solid ratio of 10.l/kg and, in case of granular wastes (size < 4 mm), has a duration of 24 h without leachant renewal. The release results have been previously reported together with the compulsory limits for landfilling of both hazardous and non-hazardous wastes (D.M. 27/09/2010) [48]. The authors reported that MSWI-FA must be stabilized in order to reduce the release of some heavy metals and, in addition, of chlorides. In fact, even if the final option is landfilling, the disposal is not allowed because sometimes both the hazardous and non-hazardous limits are exceeded. Specifically, the release of cadmium exceeds the two limits in the case of ash A and B. Chromium release exceeds the landfill disposal limit for non-hazardous wastes in the cases of all the ash, while lead release exceeds both the limits in the case of ash B and only the limit for non-hazardous wastes disposal in the cases of the other two ash. All of the three ash showed values of zinc release slightly higher than the two limits. In the case of chloride release, the values were always much higher than the two limits, while only in the case of ash B, the sulfate release slightly exceeded the limits.

Furthermore, the need for an optimized washing pretreatment of ash must also be explored to improve stabilization/solidification (S/S) process efficiency. In fact, to address an economically sound proposal, the waste amounts have to be maximized and, as a consequence, a preliminary washing treatment of ash could be worthy of consideration. Following the results reported by Colangelo et al. [17] in the above cited experimentation, a double step washing treatment with a water to ash ratio of 2:1 has been applied to the present work where a geopolymeric stabilizing system is studied. This kind of process, although more complex, minimizes water consumption, and therefore, contributing to the economy of the whole process. In fact, each ash sample has been divided into equal parts, and in the first step, one of them has been washed with a water to solid ratio of 4:1. During the second step, the solution coming from the first step is contacted with the other part of the ash. In this way, each part of ash is in contact with an amount of water useful for better handling of the liquid/solid suspension. Moreover, the overall liquid/solid ratio is 2:1. Geopolymer systems produced from coal fly ash have been used as stabilizing matrices of the three different MSWI-FA. The coal fly ash used for the synthesis of the geopolymers has been supplied by the Italian electricity board (ENEL S.p.A., Rome, Italy) and comes from a power plant located in Brindisi (Southern Italy). It is the same as that used in a previous work [21] and its characterization, made by means of the same chemical analytical techniques as reported for MSWI-FAs, has given the following chemical composition: SiO_2, 44.3% (2060 mg,Si/kg,FA) Al_2O_3, 20.2% (1070 mg/kg); Fe_2O_3, 10.5% (734 mg/kg); K_2O, 8.1% (737 mg/kg); CaO, 0.5% (36 mg/kg); Na_2O, 0.3% (22 mg/kg); loss on ignition at 1050 °C, 11.3%. Alkali activation, necessary to promote polycondensation, has been carried out by adding NaOH and sodium silicate solutions of proper concentration.

The three samples of MSWI-FA have been submitted to the stabilization treatment both as received, and after partial soluble salt removal (mainly chlorides and sulfates) carried out by the double step water washing previously described.

The compositions of the systems tested are reported in Table 2 and have been designed by fixing at 75/25 the MSWI-FA/coal fly ash ratio. Cylindrical samples (diameter 3 cm, height 6 cm) have been prepared by pouring each mixture into polyethylene moulds. Three samples have been cured for three days at 60 °C in oven under 100% relative humidity (RH) conditions. Afterwards, the specimens have been extracted from the moulds and subjected to Unconfined Compressive Strength (UCS) determination by using a 100 kN capacity Controls® MCC8 testing machine.

This mechanical evaluation is significant because it is well known that rapid strength development is a peculiar feature of geopolymerization. It is important to underline that in previous works the MSWI-FA/solid precursor ratios were much lower than 75/25. Specifically, Lancellotti et al. [40] employed systems based on metakaolin/MSWI-FA mixtures with about

17% ash, while Luna Galiano et al. [41] used about 26% ash in respect to coal fly ash in geopolymeric systems based on coal fly ash/MSWI-FA, coal fly ash + blast furnace slag/MSWI-FA, coal fly ash + metakaolin/ MSWI-FA and coal fly ash + kaolin/MSWI-FA.

The complete set of experimental compositions is reported in Table 1. These compositions have been designed taking into account those studied in the previous work [21] that gave good geopolymerization results and also by considering that a large portion of coal FA is replaced by MSWI-FA in this work. The components of all the systems listed in Table 2 have been carefully mixed and the resulting mixtures have been kept in small polyethylene cylinders of size d ×h = 3 cm ×6 cm. The polycondensation reaction has been carried out at 25 °C for times equal to 1, 3, 7, 14 and 28 days.

TABLE 1: Composition of the geopolymer materials, wt %.

System	MSWI-FA	Coal fly ash	Sodium silicate solution (1.15 M)	NaOH solution
GA_{AR}[1]	48	16	18	18 (10 M)
GA_{W}[2]	51.5	16.5	16	16 (10 M)
GB_{AR}	57.5	18.5	12	12 (10 M)
GB_{W}	60	20	10.5	10.5 (10 M)
GC_{AR}	53	17	15	15 (17 M)
GC_{W}	55	18	13.5	13.5 (17 M)

Notes: [1]*GX_{AR}: geopolymer mixture containing MSWI-FA type X (X = A or B or C) as received;* [2]*GX_{w}: geopolymer mixture containing MSWI-FA type X (X = A or B or C) prewashed.*

The specimens obtained at any prefixed polycondensation time have been characterized by means of a Thermo Scientific Nicolet Nexus FT-IR spectrometer (Thermo Scientific, Waltham, MA, USA) equipped with a DTGS KBr (deuterated triglycine sulfate with potassium bromide windows) detector. FT-IR absorption spectra have been recorded in the 4000–400 cm^{-1} range. A spectral resolution of 2 cm^{-1} has been chosen. 2.0 mg of each test sample has been mixed with 200 mg of KBr in an agate mortar, and then pressed into 200 mg pellets of 13 mm diameter. The spectrum of each sample represents an average of 32 scans. Furthermore, a Philips PW 1730 X-ray diffractometer (Philips, Eindhoven, The Netherlands) (CuKα radiation, 40 kV, 40 mA, 2θ range from 10°to 80°, equivalent step size 0.0179° 2θ, equivalent counting time 120 s per step) has been employed in order to obtain the mineralogical characterization of the same series of samples. Selected hardened samples have been also submitted to a microstructural characterization by means of a FEI Quanta 200 FEG scanning electron microscope (FEI, Hillsboro, OR, USA).

The same specimens have been used for the quantitative determination of water and sodium silicate consumed during the polycondensation reaction. The amounts of reacted sodium silicate and water at any polycondensation time have been determined as follows. Each specimen has been ground under acetone, filtered and washed with diethyl ether to remove all the residual aqueous phase. Finally, the samples have been heated in an oven up to 40 °C in order to ensure the loss of any residual fraction of the liquids previously used. The cumulative amount of reacted sodium silicate and water has been obtained by weight difference between the solid recovered after the above treatments and the ash initially employed. The amount of reacted water has been determined by the excess loss on ignition of the recovered solid over that of the initial ash. This method is extensively described in previous works [21,35].

The leaching behaviour of the stabilized systems has been assessed submitting cubic specimens of 4 cm in size to UNI 10802 test (UNI 10802, 2004) [13]. This procedure follows the protocol for monolithic specimens, which imposes water renewals after 2 and 18 h, for a total duration of 48 h. The solid surface to liquid ratio has been fixed at 1:10. At the end of each test, the pH of leachate has been measured.

To evaluate the suitability of the stabilized/solidified geopolymeric systems containing pre-washed ash for material reuse, three series of cubic specimens of 4 cm in size have been cured for 28 days at room temperature and 100% RH. Then, they have been submitted to density evaluation and UCS measurements making use of the same testing machine described above.

TABLE 2: Chemical composition of municipal solid waste incinerators (MSWI) fly ash, mg/kg.

Component	Samples		
	A	B	C
Ca	230,000 ± 11,200	270,000 ± 11,700	165,000 ± 9,900
Cl^-	113,000 ± 8,100	49,000 ± 3,700	75,000 ± 4,800
Si	110,000 ± 1,800	130,000 ± 2,200	97,000 ± 2,100
SO_4^{2-}	29,000 ± 5,500	68,000 ± 11,500	34,000 ± 7,200
Na	15,000 ± 1,050	28,000 ± 1,530	119,000 ± 9,800
Fe	12,000 ± 2,300	10,700 ± 2,040	9,450 ± 1,790
Al	12,000 ± 380	27,000 ± 1,050	14,000 ± 440
K	11,300 ± 1,080	17,000 ± 1,140	24,000 ± 1,840
Zn	9,100 ± 530	6,230 ± 430	8,400 ± 440
Pb	8,950 ± 460	17,110 ± 980	6,580 ± 270
Mg	8,500 ± 210	7,500 ± 190	1,240 ± 40
Cu	815 ± 59	6,220 ± 390	4,114 ± 220
Ni	130 ± 6	163 ± 8	117 ± 6
Ba	112 ± 22	227 ± 43	185 ± 37
Cr_{tot}	85 ± 24	270 ± 75	412 ± 95
Cd	65 ± 13	217 ± 41	88 ± 15
As	4.2 ± 1.3	5.9 ± 1.7	2.1 ± 0.9

7.3 RESULTS AND DISCUSSION

7.3.1 MSWI-FA CHEMICAL CHARACTERIZATION AND GEOPOLYMERIZATION

The results of the chemical analysis are shown in Table 2. In this work, an evaluation of the most abundant oxides was also made. The reported results revealed amounts of CaO, SiO_2 and Al_2O_3 in the ranges 23.0–32.2, 15.2–20.4 and 5.2–10.2 wt %, respectively. These data have been taken into account for the formulation of the stabilizing geopolymeric systems proposed in the present work.

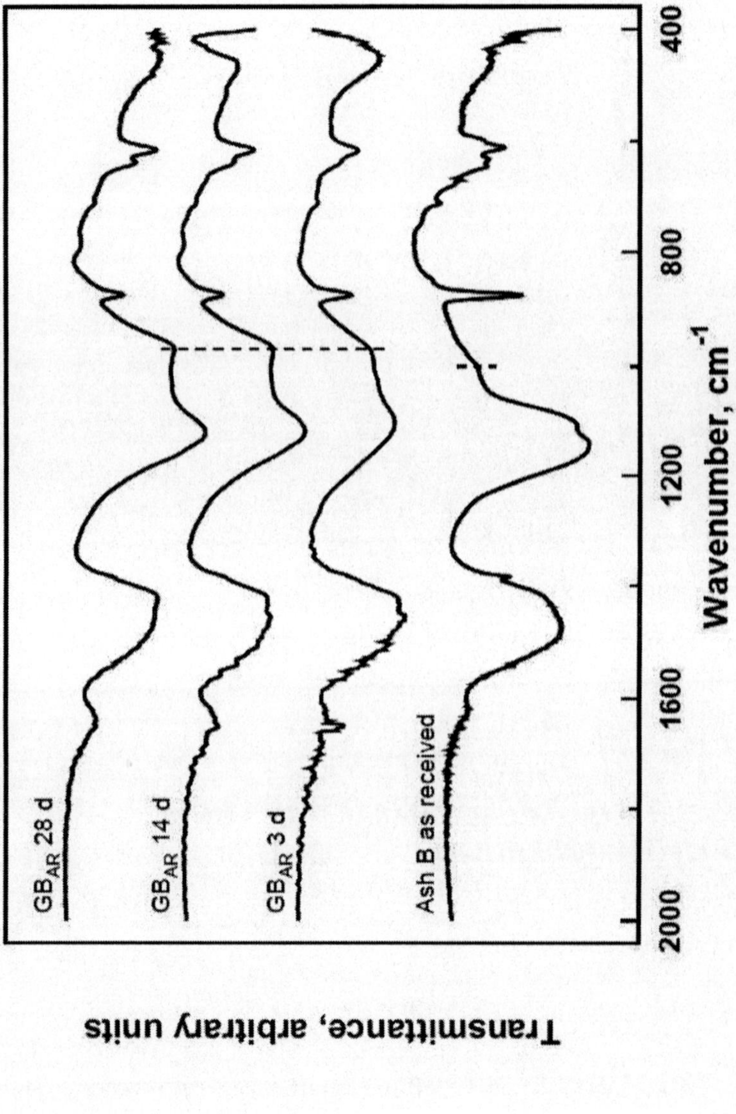

FIGURE 1: FT-IR characterization of system GBAR at several selected polycondensation times.

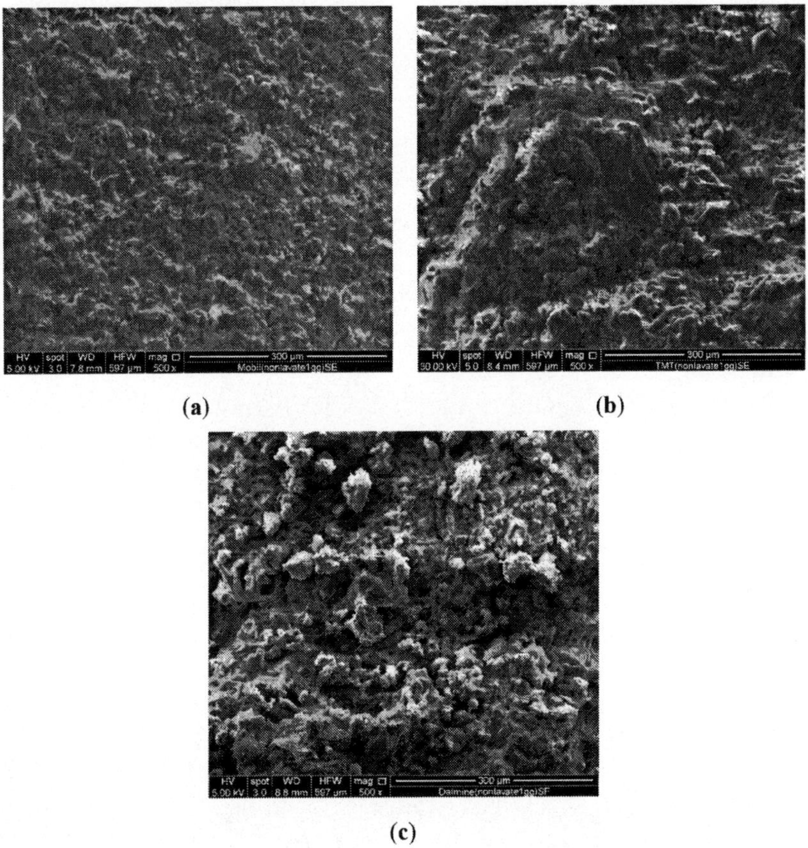

FIGURE 2: SEM Micrographs of systems (a) GAAR 28 days; (b) GBAR 28 days and (c) GCAR 28 days.

Infrared spectroscopy is a useful tool for revealing the formation of geopolymers. In fact, in FT-IR traces of raw silicates and silico-aluminates, the Si–O asymmetric stretching in tetrahedra is responsible for an absorption band centred at about 1000 cm^{-1}. When geopolymers are formed, this band is shifted to lower wavenumbers as a consequence of polycondensation with alternating Si–O and Al–O bonds (see dashed lines in Figure 1). This phenomenon can be clearly seen in Figure 1, where the results of FT-IR characterization are reported for system GBAR at some selected polycondensation times.

The band originally present at 1032 cm^{-1} in the trace relative to ash B as received shifts to 977 cm^{-1} after a polycondensation time of 28 days. In addition, the intensity of this band increases with time, indicating a corresponding increase of polycondensation degree. Figure 2 shows the micrographs of the three systems investigated after 28 days of curing. The three systems contain the ash A(a), B(b) and C(c) as received (i.e., without partial soluble salts removal).

Despite the fact that the results of FT-IR investigation show that polycondensation takes place in all the systems, the morphology of the cured samples containing the ash as received does not appear so compact to favour the development of good physico-mechanical properties. This observation holds for all the ash, even if the content of soluble salts is quite different from case to case. Figure 3 shows the micrographs of the system containing A ash previously washed and cured for 28 days (the same time considered for the unwashed systems of Figure 2).

In this case, the specimens morphology looks more compact (Figure 3a), able to favour higher strength values. Furthermore, Figure 3b shows that polycondensation actually takes place; in fact, amorphous N-A-S-H gel-phase produced during the reaction grow on the reactive coal fly ash particles. Figure 4 shows the X-Ray diffraction patterns of sample BAR and of samples GBAR after 3, 7 and 28 days of curing.

All patterns show several crystallographic peaks in a substantially amorphous matrix. The major crystalline phases identified in the sample BAR are Halite (NaCl, JCPDS card No. 5-628), Calcite ($CaCO_3$, JCPDS card No. 5-586), Anhydrite ($CaSO_4$, JCPDS card No. 37-1496), together with a low amount of Quartz (SiO_2, JCPDS card No. 46-1045) and Anorthite ($CaAl_2Si_2O_8$, JCPDS card No. 41-1486). Mineralogical composition of the fly ash is in good agreement with literature data [49]. Geopolymerized GBAR samples substantially contain the same crystalline phases as the BAR sample. Noteworthy is the disappearance of the Anhydrite and the appearance of Thenardite (Na_2SO_4, JCPDS card No. 37-1465), almost certainly due to the addition of a high amount of Na in the geopolymerization process. The distinguishing feature of the diffractogram of any geopolymer is a broad "hump" centered at approximately 27°–29° 2θ [50]. In Figure 4, a slight increase of this hump can be observed in the geopolymerized samples in respect to the as-received MSWI-FA. The same considerations can be made analysing the XRD patterns of the other two MSWI-FA samples employed.

The effect of pre-washing on the crystalline phase content of MSWI-FA is very similar to that observed by other authors [46]. Water washing determines the disappearance or decrease in intensity of the chlorides and sulfates containing phases. The XRD patterns (data not shown) of geopolymeric specimens containing pre-washed MSWI-FA do not show significant differences in terms of the above cited amorphous hump aspect.

The quantitative data of reacted water and silicate are reported in Tables 3 and 4 for all the systems studied and at all the polycondensation times investigated. The data of Table 3 show that the amount of water bound to the geopolymers decreases as the polycondensation time increases. This is a direct consequence of the reaction mechanism: initially, the starting materials dissolve in the highly alkaline reaction medium giving rise to the formation of geopolymer precursors in which several hydroxyl groups are present; then, crosslinking of these precursors takes place and the polycondensation occurs with water expulsion [51].

TABLE 3: Amount of reacted water in mg/g of initial ash.

System	Polycondensation time (days)				
	1	3	7	14	28
GA_{AR}	65.6	51.3	51.7	49.9	37.9
GA_W	83.8	59.7	46.6	43.5	36.5
GB_{AR}	72.3	41.3	51.7	9.9	0.7
GB_W	59.8	59.7	46.6	32.5	24.5
GC_{AR}	17.9	41.3	21.3	19.9	12.3
GC_W	108.4	63.7	54.6	62.5	64.5

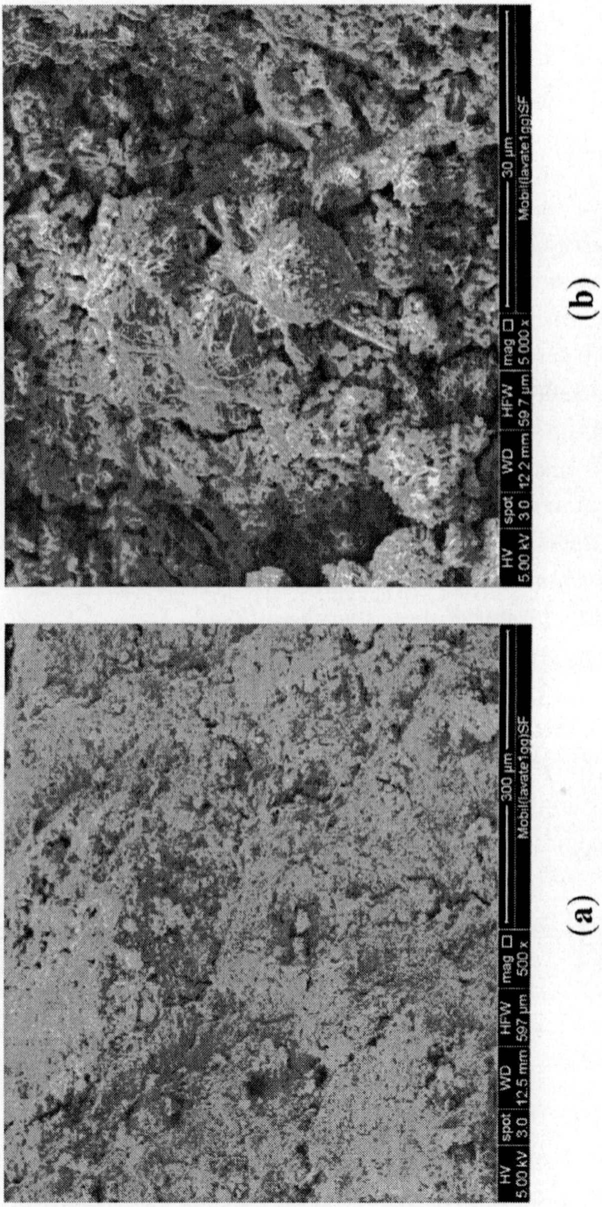

FIGURE 3: SEM Micrographs of system containing washed A ash, (a) 500× and (b) 5000×magnifications.

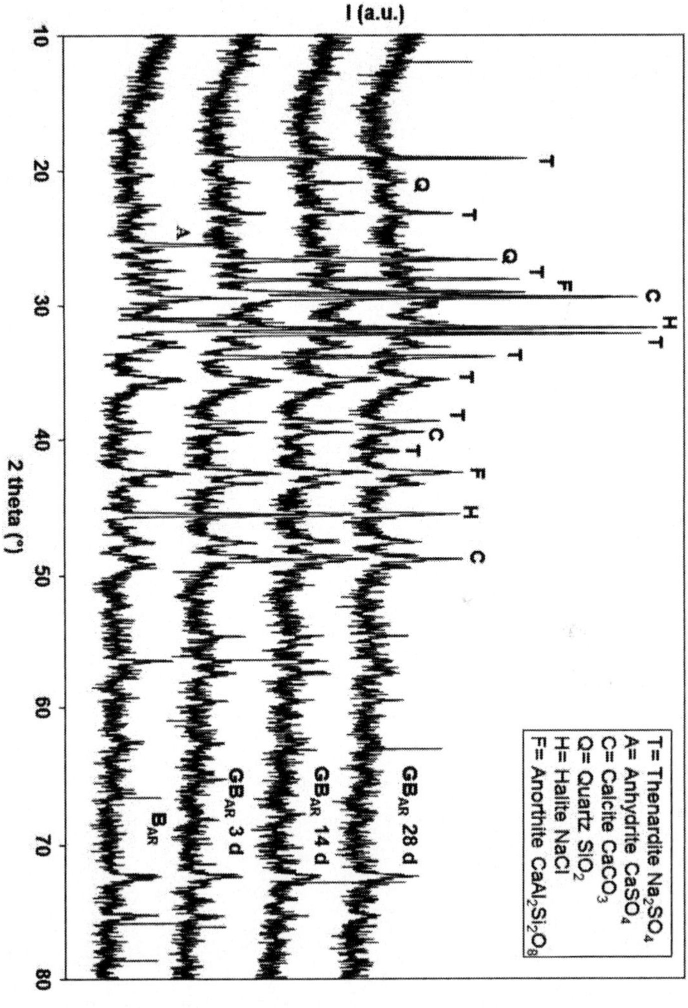

FIGURE 4: X-ray diffraction (XRD) patterns of BAR, GBAR 3 days, GBAR 14 days and GBAR 28 days samples.

TABLE 4: Amount of reacted silicate in mg/g of initial ash.

System	Polycondensation time (days)				
	1	3	7	14	28
GA_{AR}	91.8	109.2	105.2	111.9	120.8
GA_{W}	92.3	98.8	104.8	112.8	113.9
GB_{AR}	47.4	68.2	70.2	78.9	100.8
GB_{W}	69.3	69.8	74.8	82.8	82.9
GC_{AR}	47.4	69.2	30.2	78.9	60.8
GC_{W}	89.3	130.4	100.7	102.8	65.9

The data of Table 4 show that, despite a few exceptions, the amount of reacted silicate increases with reaction time. The quantitative data of Table 4 can be compared with similar results obtained in the previous work [21] in which coal FA was used on its own. In the above cited work, it was found that starting with SiO_2/Al_2O_3 ratios equal to 4 and 6, the amount of reacted silicate at 25 °C and after 28 days, reached the values of about 200 and 250 g/g of initial FA, respectively. In this work, the presence of MSWI-FA worsens these results, but not dramatically, inasmuch as values ranging from about 60–120 mg/g are reached under the same experimental conditions (see Table 4). Some differences can be seen in relation to MSWI-FA origin, but in all the cases, the degree of polycondensation is high enough to get monolithic products. Reducing the content of chlorides by washing has a minor effect, if any. This is particularly relevant in relation to MSWI-FA stabilization, as it is well known that the effectiveness of traditional cement-based matrices can be severely compromised by high chloride content.

7.3.2 MSWI-FA STABILIZATION FOR SAFER DISPOSAL

The results of the leaching tests carried out on the stabilized specimens containing MSWI-FA as-received are reported in Table 5. If these results are compared with those reported by Colangelo et al. in a previous work [17], it can be seen that the geopolymer-based stabilizing system is more efficient in respect to a cement-based one. The values reported in parentheses are relative to systems where an 80/20 MSWI-FA to cement ratio has been imposed and the ash have been mixed without a washing pre-treatment. The authors found that the process had a limited positive effect on the leaching behavior of chlorides and sulfates.

TABLE 5: Results of UNI 10802 leaching test on stabilized systems containing MSWI-FA as-received, mg/L.

Components	System			Limits for non-hazardous wastes
	GA_{AR}	GB_{AR}	GC_{AR}	
A_s	<0.10 (0.10)	<0.10 (0.12)	<0.10 (<0.10)	0.2
Ba	0.17 (0.34)	<0.10 (<0.1)	0.31 (0.57)	10
Cd	<0.10 (0.18)	<0.10 (0.33)	<0.10 (<0.10)	0.1
Cr_{tot}	0.91 (1.31)	0.67 (0.91)	0.45 (0.98)	1
Ni	0.18 (0.53)	0.21 (0.98)	0.14 (0.75)	1
Pb	0.52 (1.31)	1.14 (1.52)	0.50 (0.91)	1
Cu	0.10 (0.15)	1.18 (4.18)	0.43 (1.01)	5
Zn	1.64 (1.69)	1.12 (1.82)	0.87 (0.97)	5
Cl^-	5080 (7140)	2115 (3015)	3450 (4780)	1500
SO_4^{2-}	1080 (1480)	3160 (4150)	1570 (1830)	2000

Note: Results of previous cement-stabilization/solidification process with MSWI-FA/cement = 80/20.

The data of Table 5 show that, although the geopolymer system is more effective in respect to pollutant release, the resulting values for chlorides are still higher than the limits imposed by Italian regulation (D.M. 27/09/2010, 2010) [48] for disposal of stabilized wastes in landfill for non-hazardous wastes. The improvement of effectiveness is partially relevant for chlorides, as the release is reduced by 29%, 30% and 28% in the case of systems GAAR, GBAR and GCAR, respectively. Despite the better immobilization compared to what was found by Colangelo et al. [17], release of sulfates and lead even exceeds the above limits for system

GBAR. If these results are compared with those obtained by other authors on geopolymer-based stabilizing systems, including MSWI-FA, it is possible to see that the most important findings are in agreement. Lancellotti et al. [40] studied the stabilization of two MSWI-FA samples by employing a metakaolin-based geopolymer matrix. In the cited work, the wastes were stabilized without a specific washing pretreatment with a metakaolin/MSWI-FA ratio of 5/1. After curing, the leaching behavior and the chemical stability of the matrix were assessed showing that the systems could be disposed of in a landfill for non-hazardous wastes. Luna Galiano et al. [41] used systems containing coal fly ash, blast furnace slag, metakaolin and kaolin in different ratios to stabilize an unwashed MSWI-FA sample. A comparison with ordinary Portland cement and lime-based systems was made through compressive strength and leaching behavior evaluation. Zheng et al. [43] employed coal fly ash-based geopolymer binders to evaluate both the effect of Si/Al ratio and alkali content on heavy metal release and the microstructure of systems containing untreated MSWI-FA. Particularly, Lancellotti et al. [40], in agreement with others researches [38,39,41–43], found that cadmium is highly immobilized in geopolymer matrices due to the very low solubility of $Cd(OH)_2$ in the highly alkaline leachate of the coal fly ash-based geopolymer system. In addition, the leaching behavior of nickel, chromium, copper and lead is also comparable.

As in this case, Luna Galiano et al. [41] reported comparisons between the stabilizing efficiency of various coal fly ash/geopolymer and cementitious mixtures. They found similar differences in leaching behavior between the two different binding systems. As far as the chloride release is concerned, the values detected in our systems prove that a pre-washed treatment of fly ash is required. Table 6 shows the release values of the stabilized geopolymer-based systems containing the MSWI-FA after the two-step washing treatment.

It can be seen that, as in the previous study of the same fly ash [17], the chloride extraction strongly reduces the release values measured according to the UNI 10802 [13] leaching test. The measured values are lower than the limits fixed for disposal of stabilized (unreactive) wastes. As expected, all the heavy metal release values decrease below the limits and consequently, the stabilized systems are suited for safer disposal if landfilling is the final disposal option.

TABLE 6: Results of UNI 10802 leaching test on stabilized systems containing two-step 2:1 washed MSWI-FA, mg/L.

Components	System			Limits for non-hazardous wastes
	GA_W	GB_W	GC_W	
As	<0.10 (0.10)	<0.10 (0.12)	<0.10 (<0.10)	0.2
Ba	0.11 (0.31)	<0.10 (<0.10)	0.28 (0.38)	10
Cd	<0.10 (0.17)	<0.10 (0.23)	<0.10 (<0.10)	0.1
Cr_{tot}	0.80 (1.10)	0.53 (0.80)	0.43 (0.53)	1
Ni	<0.10 (0.50)	0.16 (0.73)	<0.10 (0.58)	1
Pb	0.47 (1.47)	1.05 (1.34)	0.31 (0.81)	1
Cu	0.10 (0.17)	0.71 (3.71)	0.53 (0.93)	5
Zn	1.61 (1.71)	1.07 (1.57)	0.81 (0.91)	5
Cl^-	1240 (1840)	1410 (910)	1160 (1170)	1500
SO_4^{2-}	480 (630)	550 (450)	580 (460)	2000

Note: Results of previous cement-stabilization/solidification process with MSWI-FA/cement = 80/20.

The economic advantages of this final option are evident considering the possibility to dispose of the MSWI-FA in a less expensive landfill for non-hazardous wastes. It is so because the overall water requirement is limited, even if the two-step pre-treatment seems to be more complex.

In all the leaching steps, the pH has been measured and the detected values were always highly alkaline (>11). This is strongly associated to the nature of the mixtures and agrees with previous findings on similar systems where different MSWI-FA have been mixed in various ratios with coal fly ash, metakaolin, kaolin and blast furnace slag [38–43].

7.3.3 MSWI-FA STABILIZATION/SOLIDIFICATION AND MANAGEMENT FOR MATERIAL REUSE

The results of UCS measurements on the three hardened geopolymeric systems, after 7 and 28-day curing, are shown in Figure 5 together with the values previously measured on cement-based solidified systems containing MSWI-FA cured for 28 days and proposed as bound granular material for road basement [17].

In consideration of the high fire-resistance and hardness of geopolymeric materials, the experiments were carried out on hardened specimens in order to achieve a final product in the field of decorative or non-structural applications, such as brick fireplaces, hearths, patios, etc.

The manufacturing of geopolymer bricks (geobricks) based on alkali activated coal fly ash-based systems was already studied by Palomo et al. [52]. They explored possible applications of geobricks as monoblock, lightweight matrices and fire resistant tiles where geopolymer typical technological properties can be advantageously exploited.

In other studies, Ariöz et al. [53,54] produced geobricks making use of coal fly ash, sodium hydroxide and sodium silicate solution curing the mixtures at up to 75 °C and in the presence of forming pressure. Furthermore, a number of commercial manufacturing processes have already been developed showing interesting market areas for the final products [55,56]. In all the cited cases, the role of both curing temperature and forming pressure were underlined in terms of strength, density, porosity, heat conductivity, etc. The results showed that also at low curing temperature, the systems investigated gave a technical performance adequate for a wide range of applications. It can be seen that the data of geopolymer type systems cured at room temperature are lower than that of cementitious ones. In particular, the mechanical performances of all the type G mixtures are very similar with the only exception for the system containing type B MSWI-FA. In fact, the higher chlorides residual content present in ash B gives, on the corresponding system, a slightly lower compressive strength than that shown by the other two mixtures. This is in agreement with the findings of the microstructural characterization. The values of the density measured on specimens cured for 28 days are: 1420, 1340 and 1435 kg/m^3 for the systems GAw, GBw and GCw, respectively. Also, in this case, a comparison with cement-based systems is carried out. The trend of the results agrees with that observed in Figure 5.

All the measured absolute physico-mechanical values are quite close to that of soft masonry stones like clay or zeolite-based Neapolitan yellow tuff bricks. The mean compressive strength of the latter type of stones ranges between 2 and 5.73 MPa, while the density values are in the range of 1500–1650 kg/m^3 , as reported in a recent wide investigation on compressive behavior of tuff masonry panels [57]. The entire set of physical and mechanical data shows that, in line with this view, the stabilization with 25% of binder is technologically sound for the proposed way of material reuse.

In Italy, specific authorization is needed before material reuse can be put into practice by means of hazardous waste stabilization processes. The MSWI-FA stabilization process studied in this work can only be considered an economically interesting proposal and a feasible technique for material reuse in the field of backfilling in abandoned quarries. To this regard, due to the specific geology of many areas present in Campania Region and the consequent huge amounts of quarried tuff stones, it can easily be evaluated that the proposed application could absorb very high quantities of stabilized MSWI-FA.

As far as the process economy is concerned, considerations similar to those presented in the previous paper, where cement stabilization was proposed, can be made. In that case, the complete ash treatment process cost (together with the washing-salt disposal) was estimated to be cheaper than the non-hazardous landfill disposal [17]. In the case of geopolymer stabilization process, this alternative treatment of MSWI-FA could be considered less expensive and more environmentally friendly. In fact, even if the cost of geopolymer matrix components is not yet standardized, the possibility to employ industrial solid waste, such as coal fly ash, is a very attractive option.

7.4 CONCLUSIONS

This work has proved that mixtures containing coal fly ash and pre-washed MSWI-FA can be employed for the synthesis of geopolymeric systems. Three different samples of MSWI-FA have been used and in all the cases the polycondensation took place with formation of monolithic products. The experiments have been also carried out with MSWI-FA in which the soluble salt content had been significantly lowered by a water washing process optimized in relation to water consumption.

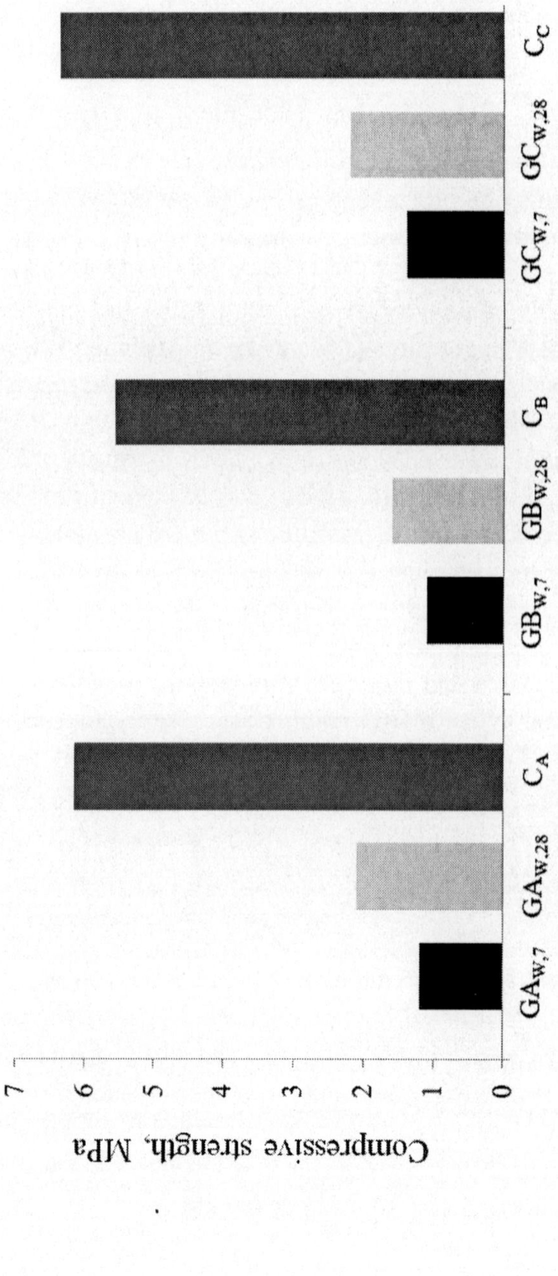

FIGURE 5: Compressive strength of MSWI-FA geopolymeric cubic specimens.

Leaching tests have been carried out on both as-received and washed MSWI-FA, showing that the geopolymer-based matrix has a better stabilizing effect in comparison to previously studied cementitious systems. Chemical and microscopic analyses proved that the content of soluble salts plays a minor role in the amounts of reacted water and silicate, but strongly affects the microstructure of the neo-formed phases.

Finally, it can be argued that, considering the results of the physico-mechanical tests, MSWI-FA washing could be very advantageous from the point of view of safer ash disposal and their recycling as backfilling blocks for abandoned quarries or low temperature setting geopolymer soft bricks.

REFERENCES

1. Aubert, J.E.; Husson, B.; Sarramone, N. Utilization of municipal solid waste incineration (MSWI) fly ash in blended cement: Part 1: Processing and characterization of MSWI fly ash. J. Hazard. Mater. 2006, 136, 624–631.
2. Cioffi, R.; Colangelo, F.; Montagnaro, F.; Santoro, L. Manufacture of artificial aggregate using MSWI bottom ash. Waste Manag. 2011, 31, 281–288.
3. Hjelmar, O. Disposal strategies for municipal solid waste incineration residues. J. Hazard. Mater. 1996, 47, 345–368.
4. Andini, S.; Cioffi, R.; Colangelo, F.; Montagnaro, F.; Santoro, L. Effect of mechanochemical processing on adsorptive properties of blast furnace slag. J. Environ. Eng. 2013, doi:10.1061/(ASCE)EE.1943-7870.0000768.
5. Beretka, J.; Cioffi, R.; Santoro, L.; Valenti, G. Cementitious mixtures containing industrial process wastes suitable for the manufacture of preformed building elements. J. Chem. Technol. Biotechnol. 1994, 59, 243–247.
6. Cioffi, R.; Maffucci, L.; Martone, G.; Santoro, L. Feasibility of manufacturing building materials by recycling a waste from ion exchange process. Environ. Technol. 1998, 19, 1145–1150.
7. Colangelo, F.; Cioffi, R.; Lavorgna, M.; Verdolotti, L.; De Stefano, L. Treatment and recycling of asbestos-cement containing waste. J. Hazard. Mater. 2011, 195, 391–397.
8. Colangelo, F.; Cioffi, R. Use of cement kiln dust, blast furnace slag and marble sludge in the manufacture of sustainable artificial aggregates by means of cold bonding pelletization. Materials 2013, 6, 3139–3159.
9. Iucolano, F.; Liguori, B.; Caputo, D.; Colangelo, F.; Cioffi, R. Recycled plastic aggregate in mortars composition: effect on physical and mechanical properties. Mater. Des. 2013, 52, 916–922.
10. Limbachiya, M.C. Bulk engineering and durability properties of washed glass sand concrete. Constr. Build. Mater. 2009, 23, 1078–1083.

11. Poon, C.S.; Kou, S.C.; Lam, L. Use of recycled aggregates in molded concrete bricks and blocks. Constr. Build. Mater. 2002, 16, 281–289.
12. Sani, D.; Moriconi, G.; Fava, G.; Corinaldesi, V. Leaching and mechanical behaviour of concrete manufactured with recycled aggregates. Waste Manag. 2005, 25, 177–182.
13. Wastes—Liquid, Granular Wastes and Sludges—Manual Sampling and Preparation and Analysis of Eluates; UNI 10802: 2004; Italian Standards: Rome, Italy, 2004.
14. Characterisation of Waste—Leaching—Compliance Test for Leaching of Granular Waste Materials and Sludges; EN 12457-2: 2002; British Standards Institution: London, UK, 2002.
15. Albino, V.; Cioffi, R.; Pansini, M.; Colella, C. Disposal of lead-containing zeolite sludges in cement matrix. Environ. Technol. 1995, 16, 147–156.
16. Cioffi, R.; Pansini, M.; Caputo, D.; Colella, C. Evaluation of mechanical and leaching properties of cement-based solidified materials encapsulating Cd-exchanged natural zeolites. Environ. Technol. 1996, 17, 1215–1224.
17. Colangelo, F.; Cioffi, R.; Montagnaro, F.; Santoro, L. Soluble salt removal from MSWI fly ash and its stabilization for safer disposal and recovery as road basement material. Waste Manag. 2012, 32, 1179–1185.
18. Mangialardi, T. Disposal of MSWI fly ash through a combined washing-immobilisation process. J. Hazard. Mater. 2003, 98, 225–240.
19. Albino, V.; Cioffi, R.; De Vito, B.; Santoro, L. Evaluation of solid waste stabilization processes by means of leaching tests. Environ. Technol. 1996, 17, 309–315.
20. Taylor, H.F.W. Cement Chemistry; Thomas Telford Publishing: London, UK, 1997.
21. Andini, S.; Cioffi, R.; Colangelo, F.; Grieco, T.; Montagnaro, F.; Santoro, L. Coal fly ash as raw material for the manufacture of geopolymer-based products. Waste Manag. 2008, 28, 416–423.
22. Davidovits, J. Geopolymer, Chemistry and Applications, 3rd ed.; Institute Geopolymere: Saint-Quentin, France, 2011; pp. 10–11.
23. Palomo, A.; Macias, A.; Blanco, M.T.; Puertas, F. Physical, Chemical and Mechanical Characterization of Geopolymers. In Proceedings of the 9th International Congress on the Chemistry of Cement, New Delhi, India; National Council for Cement and Building Materials (NCCBM): New Delhi, India, 1992; pp. 505–511.
24. Palomo, A.; Blanco-Varela, M.T.; Granizo, M.L.; Puertas, F.; Vasquez, T.; Grutzeck, M.W. Chemical stability of cementitious materials based on metakaolin. Cem. Concr. Res. 1999, 29, 997–1004.
25. Schmücker, M.; MacKenzie, K.J.D. Microstructure of sodium polysialate siloxo geopolymer. Ceram. Int. 2005, 31, 433–437.
26. Bernal, S.A.; Mejía de Gutiérrez, R.; Pedraza, A.L.; Provis, J.L.; Rodriguez, E.D.; Delvasto, S. Effect of binder content on the performance of alkali-activated slag concretes. Cem. Concr. Res. 2011, 41, 1–8.
27. Colangelo, F.; Roviello, G.; Ricciotti, L.; Ferone, C.; Cioffi, R. Preparation and characterization of new geopolymer-epoxy resin hybrid mortars. Materials 2013, 6, 2989–3006.
28. Ferone, C.; Colangelo, F.; Cioffi, R.; Montagnaro, F.; Santoro, L. Mechanical performances of weathered coal fly ash based geopolymer bricks. Procedia Eng. 2011, 21, 745–752.

29. Ferone, C.; Colangelo, F.; Cioffi, R.; Montagnaro, F.; Santoro, L. Use of reservoir clay sediments as raw materials for geopolymer binders. Adv. Appl. Ceram. 2013, 112, 184–189.
30. Ferone, C.; Colangelo, F.; Roviello, G.; Asprone, D.; Menna, C.; Balsamo, A.; Prota, A.; Cioffi, R.; Manfredi, G. Application-oriented chemical optimization of a metakaolin based geopolymer. Materials 2013, 6, 1920–1939.
31. Ferone, C.; Roviello, G.; Colangelo, F.; Cioffi, R.; Tarallo, O. Novel hybrid organic geopolymer materials. Appl. Clay Sci. 2013, 73, 42–50.
32. Menna, C.; Asprone, D.; Ferone, C.; Colangelo, F.; Balsamo, A.; Prota, A.; Cioffi, R.; Manfredi, G. Use of geopolymers for composite external reinforcement of RC members. Compos. Part B Eng. 2012, 45, 1667–1676.
33. Kong, D.L.; Sanjayan, J.G. Effect of elevated temperatures on geopolymer paste, mortar and concrete. Cem. Concr. Res. 2010, 40, 334–339.
34. Ben Haha, M.; Le Saout, G.; Winnefeld, F.; Lothenbach, B. Influence of activator type on hydration kinetics, hydrate assemblage and microstructural development of alkali activated blast-furnace slags. Cem. Concr. Res. 2011, 41, 301–310.
35. Cioffi, R.; Maffucci, L.; Santoro, L. Optimization of geopolymer synthesis by calcination and polycondensation of a kaolinitic residue. Resour. Conserv. Recycl. 2003, 40, 27–38.
36. Criado, M.; Fernández Jiménez, A.; Sobrados, I.; Palomo, A.; Sanz, J. Effect of relative humidity on the reaction products of alkali activated fly ash. J. Eur. Ceram. Soc. 2012, 32, 2799–2807.
37. Komnitsas, K.; Zaharaki D.; Perdikatsis, V. Geopolymerisation of low calcium ferronickel slags. J. Mater. Sci. 2007, 42, 3073–3082.
38. Fernández Pereira, C.; Luna, Y.; Querol, X.; Antenucci, D.; Vale, J. Waste stabilization/solidification of an electric arc furnace dust using fly ash-based geopolymers. Fuel 2009, 88, 1185–1193.
39. Komnitsas, K.; Zaharaki, D.; Bartzas, G. Effect of sulfate and nitrate anions on heavy metal immobilisation in ferronickel slag geopolymers. Appl. Clay Sci. 2012, 73, 103–109.
40. Lancellotti, I.; Kamseu, E.; Michelazzi, M.; Barbieri, L.; Corradi, A.; Leonelli, C. Chemical stability of geopolymers containing municipal solid waste incinerator fly ash. Waste Manag. 2010, 30, 673–679.
41. Luna Galiano, Y.; Fernández Pereira, C.; Vale, J. Stabilization/solidification of a municipal solid waste incineration residue using fly ash-based geopolymers. J. Hazard. Mater. 2011, 185, 373–381.
42. Van Jaarsveld, J.G.S.; Van Deventer, J.S.J.; Lorenzen, L. The potential use of geopolymeric materials to immobilise toxic metals: Part I. Theory and applications. Miner. Eng. 1997, 10, 659–669.
43. Zheng, L.; Wang, W.; Shi, Y. The effects of alkaline dosage and Si/Al ratio on the immobilization of heavy metals in municipal solid waste incineration fly ash-based geopolymer. Chemosphere 2010, 79, 665–671.
44. Lee, W.K.W.; Van Deventer, J.S.J. Effects of anions on the formation of aluminosilicate gel in geopolymers. Ind. Eng. Chem. Res. 2002, 41, 4550–4558.

45. Lee, W.K.W.; Van Deventer, J.S.J. The effects of inorganic salt contamination on the strength and durability of geopolymers. Colloids Surf. A Physicochem. Eng. Asp. 2002, 211, 115–126.
46. Zheng, L.; Wang, C.; Wang, W.; Shi, Y.; Gao, X. Immobilization of MSWI fly ash through geopolymerization: Effects of water-wash. Waste Manag.2011, 31, 311–317.
47. Standard Practice for Acid-Extraction of Elements from Sediments Using Closed Vessel Microwave Heating; ASTM D5258-02: 2013; ASTM International: West Conshohocken, PA, USA, 2013.
48. Definition of Waste Landfilling Admissibility Criteria; Italian Ministerial Decree 27/9/2010 in substitution of those contained in Ministerial Decree 3/8/2005; The Italian Minister for the Environment, Land and Sea: Roma, Italy, 2010.
49. Liu, Y.; Zheng, L.; Li, X.; Xie, S. SEM/EDS and XRD characterization of raw and washed MSWI fly ash sintered at different temperatures. J. Hazard. Mater. 2009, 162, 161–173.
50. Provis, J.L.; Lukey, G.C.; van Deventer, J.S. Do geopolymers actually contain nanocrystalline zeolites? A reexamination of existing results. Chem. Mater. 2005, 17, 3075–3085.
51. Xu, H.; Van Deventer, J.S.J. The geopolymerisation of alumino-silicate minerals. Int. J. Miner. Process. 2000, 59, 247–266.
52. Palomo, A.; Fernández-Jiménez, A. Alkaline Activation, Procedure for Transforming Fly Ash into New Materials. Part 1: Applications. In Proceedings of World of Coal Ash (WOCA) Conference, Denver, CO, USA, 9–12 May 2011.
53. Ariöz, Ö.; Kilinç, K.; Zeybek, O.; Tuncan, M.; Tuncan, A.; Kavas, T. An Experimental Investigation on Fly Ash Based Geopolymer Bricks. In Proceedings of Global Roadmap for Ceramics ICC2, Verona, Italy, 29 June–4 July 2008.
54. Ariöz, Ö.; Kilinç, K.; Zeybek, O.; Tuncan, M.; Tuncan, A.; Kavas, T. Physical and Mechanical Properties of Geobricks. In Proceedings of Third International Workshop on Advanced Ceramics (IWAC03), Limoges, France, 6–8 November 2008.
55. Geopolymeric Cross-Linking (LTGS) and Building Materials. Available online: http://www.geopolymer.org/category/library (accessed on 8 August 2013).
56. Mehrotra, S.P.; Kumar, R.; Kumar, M.B.; Kumar, S. Process for the Preparation of Self-Glazed Geopolymer Tile from Fly Ash and Blast Furnace Slag. U.S. Patent 2007/0221100, 27 September 2007.
57. Marcari, G.; Fabbrocino, G.; Lourenço, P.B. Mechanical Properties of Tuff and Calcarenite Stone Masonry Panels under Compression. In Proceedings of 8th International Masonry Conference, Dresden, Germany, 4–7 July 2010.

CHAPTER 8

A Perspective on the Prowaste Concept: Efficient Utilization of Plastic Waste through Product Design and Process Innovation

ANTONIO GRECO, MARIAENRICA FRIGIONE, ALFONSO MAFFEZZOLI, ALESSANDRO MARSEGLIA, AND ALESSANDRA PASSARO

8.1 INTRODUCTION

Although in recent years great attention has been given to the production of objects using recycled plastic materials [1,2,3,4,5], the poor quality of objects made from mixed recycled plastics, and the costs associated with processes capable of reducing impurities, place considerable constraints on the economic viability of recycling of plastics in general [6,7]. Most low density polyethylene (LDPE) coming from solid urban waste is processed by means of a process called the "in-mold extrusion", or "intrusion" process. The products obtained by this technology are used to replace wood in outdoor applications, owing to their better resistance to environmental degradation. These materials are usually referred to as "recycled plastic lumber" (RPL), and are widely used in marine and high humidity environments [8]. The poor compatibility of the different polymers present in plastic waste, together with the contamination by non-polymeric materials (above all paper), results in products with poor mechanical properties.

Usually, RPL is used for the production of high aspect ratio beams, subjected to one-directional bending forces. Under such conditions, a very efficient reinforcement of the beams can be attained by the introduction of rigid rods near the upper and lower surfaces of the beam, which can be readily achieved in continuous extrusion processes through introduction of the appropriate features in the die, and continuously feeding of the reinforcing rods [9,10,11]. Recently, it was demonstrated that a similar approach can be readily adapted to the in mold extrusion process [12], as well as to other closed mold processes, such as rotational molding [13]. The resulting product is characterized by a strong anisotropy, since the reinforcing elements are placed in the zones of the beam subjected to the higher stresses. The process has therefore been adapted for the production of reinforced RPL beams on an industrial plant, thanks to the grant Eco-Innovation promoted by EACI (European Agency for Competitiveness and Innovation). The project Prowaste (Efficient utilization of Plastic Waste through Product Design and Process Innovation) was promoted by a team of partners deeply involved in the recycling and reuse chain.

In the present work, the selective reinforcement strategy used for enhancing the mechanical properties of RPL beams, and its adaptation for the production of components on an industrial plant, is presented. A preliminary selection of different materials to be used as reinforcing elements was made, based on the evaluation of the mechanical properties, costs, weight and potential durability. Reinforced beams were then produced in an industrial plant, and characterized with respect to stiffness and creep behavior of the beam, as well as the pullout resistance of the rods from the beam. Finally, the potential advantages of the Prowaste concept compared to other more conventional processes is presented.

8.1.1 PROWASTE CONCEPT

The Prowaste concept is based on the stiffness enhancement of RPL beams by the addition of continuous rods, selectively placed at specific positions on the cross section of the beam. A possible layout of reinforcing rods for a rectangular cross section of the beam is reported in Figure 1. Reinforcement is achieved by embedding rods, characterized by a cumulative cross sectional area equal to A_{ROD}, at a distance c from the half height of the cross section of the beam. Rods are disposed symmetrically with respect to the half-height of the beam, in order to avoid thermal distortions. The flexural stiffness of the reinforced RPL beam is defined as the ratio between the applied force F and the maximum deflection at half length of the beam, v_{max}. In the case of a simply supported beam, the stiffness K is given by:

$$K = \frac{F}{v_{max}} = \frac{48E_{RPL}I_r}{L^3} \qquad (1)$$

where E_{RPL} is the elastic modulus of the RPL, and I_R is the moment of inertia of the reinforced cross section with respect to its neutral axis. The moment of inertia depends on the beam geometry, the properties of the constituent materials, and the rods layout. For the layout reported in Figure 1, it can be calculated as:

$$I_r = \frac{bh^3}{12} + nA_{ROD}c \qquad (2)$$

where n is the ratio between the modulus of the reinforcing rods, E_{ROD}, and E_{RPL}. The reinforcing efficiency of the rods increases as the distance between the rod position and the half height of the beam is increased. The moment of inertia of unreinforced RPL is simply obtained by Equation (2) neglecting the last term on the right hand side.

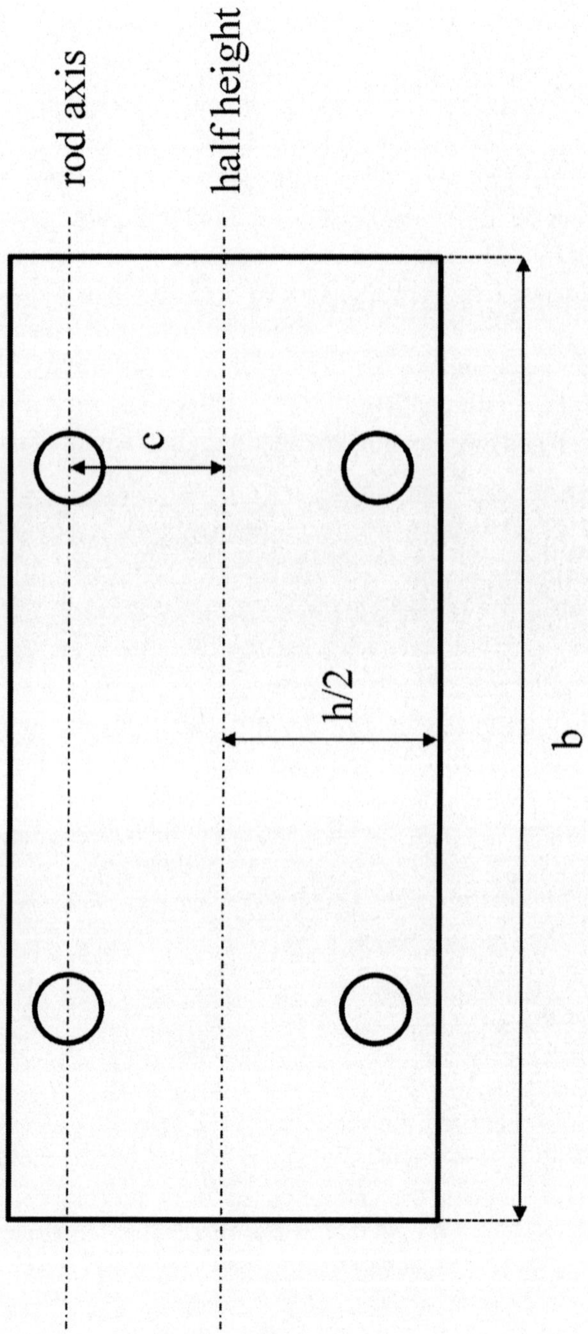

FIGURE 1: Layout of reinforcing rods.

Efficient stiffening can be attained only is a full adhesion between the RPL and reinforcement is preserved. The capability of load transfer between the RPL and the rod can be estimated by considering the shear stress at the interface:

$$\tau_{rz} I_R = 12F \frac{E_{ROD} cR}{KL^3} \quad (3)$$

where R is the radius of the rods. When the shear stress at the interface overcomes the adhesion strength, debonding at the interface causes loss of the stiffening effect.

TABLE 1: Physical properties and estimated costs of reinforcing rods.

Type of reinforcing rod	Cost (€/m) for a 3 mm Diameter Rod	Cost (€/dm³)	Flexural Modulus (GPa)	Density (Kg/dm³)
Inox steel rods	0.35	49.5	210	7.8
Aluminum rods	1.5	212	70	2.7
Glass reinforced pultruded rods based on thermosetting matrix	0.06	8.5	50	2.2
Glass reinforced pultruded rods based on thermoplastic matrix	0.3	42.5	15	1.6
Carbon reinforced pultruded rods	3	424	70	1.3

Different materials can be used to reinforce RPL beams. Among these, the materials reported in Table 1 have been evaluated. As noted in Table 1, the cost of pultruded carbon reinforced rods or aluminum rod is much higher than that of other materials, and is not compatible with the expected low cost of the products obtained from recycled plastic. The glass reinforced rods give a lower contribution to stiffness, but have a lower weight and cost compared to stainless steel. For such a reason, the pultruded glass reinforced rods were chosen to produce reinforced RPL beams. Two different types of matrices were evaluated: thermoplastic and thermoset matrices. The pultruded rods based on the thermoset matrix are characterized by an higher glass fiber content, and therefore by an higher modulus, and a lower cost compared to thermoplastic matrix based rods. Nevertheless, the higher compatibility between the thermoplastic matrix and RPL is expected to positively affect the adhesion properties, as will be better discussed in the next section.

8.1.2 PRODUCTION OF SELECTIVELY REINFORCED RPL BEAMS ON AN INDUSTRIAL PLANT

In order to incorporate the rods in the polymer mass during the in mold extrusion process, two metallic parts were properly designed and built. In the framework of the Prowaste project, Masmec SPA (Modugno, Italy) was in charge of the design of the two metallic parts. The first part is a “mask”, which is placed on the front of the mold (the surface closer to the extruder). A schematic drawing of this part is reported in Figure 2. The internal frame, depicted by horizontal hatching in Figure 2, is such that it fits inside the mold (its dimensions are $(h - 2\delta) \times (b - 2\delta)$, being δ a tolerance of about 0.5 mm). This frame has four small circular holes, which are required to position the rods, and three large square openings, which act as channels for the feeding of the polymer melt into the mold. The frame is provided with a screwing system, which allows blocking the rods in correspondence of the four small holes. The external frame, depicted by oblique hatching in Figure 2, is a flange which stops the mask from sliding forward during mold filling (its dimension are $(h + 2\Delta) \times (b + 2\Delta)$, being Δ about 10 mm). The second part is a “guide”, reported in Figure 3, which is placed inside the mold in proximity of the “mask” at the beginning of mold filling cycle and capable of sliding inside the mold as the molten polymer fills it. The outer dimension of the guide is such that is fits inside the mold (its dimensions are equal to those of the internal frame of the mask). The four holes are necessary to place the rods. Both components were made with 5 mm thick stainless steel.

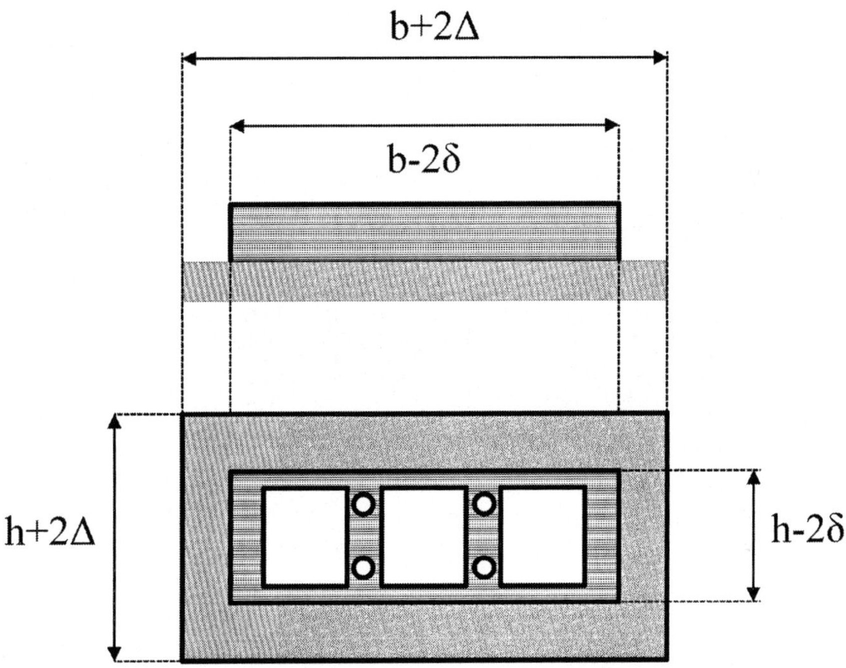

FIGURE 2: Schematic drawing of the front mask.

FIGURE 3: Schematic drawing of the sliding guide.

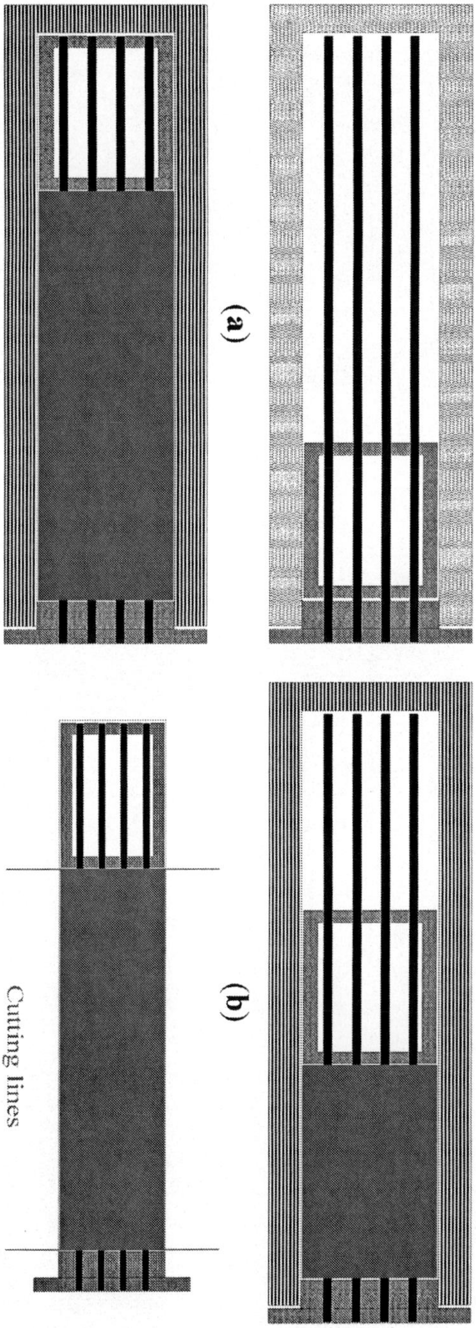

FIGURE 4: Scheme of the in mold extrusion process with selective reinforcement. (a) empty mold; (b) mold filling; (c) end of filling; (d) part extraction.

FIGURE 5: Pictures of the mask and the sliding guide.

A scheme of the process for selectively reinforcing the RPL beams is reported in Figure 4. At the beginning of mold filling the rods are introduced inside the mold, through the holes of the mask and the holes of the guide, as reported in Figure 4a. During mold filling, while the rods are held in position by the blocking system on the front surface of the mask, the guide is pushed forward by the molten polymer front. As the guide slides towards the back surface of the mold, the rods remain embedded in the polymer melt, as shown in Figure 4b. At the end of each molding cycle, the reinforced beam can be extracted, Figure 4c and the two metallic parts can be reused in the following cycle after cutting, Figure 4d. If so desired by alternative product design considerations, the layup of the rods can be changed by substituting the mask and the guide at the end of each molding cycle. As an example, in Figure 5, one picture of two masks and two sliding guides is reported. The two couples of tools are different, allowing for introduction of the rods at different distances from the half height of the beams.

In the framework of the Prowaste project, Solteco SL (Alfaro, Spain) was in charge of production of reinforced RPL. The extruder used to produce the beams is a model Kuhne 140/100, 25 L/D, equipped with forced feeding. The beams produced are 2960 mm long and 40 × 120 mm in cross section.

For each molding test, four pultruded rods were incorporated in the RPL beam, with a symmetrical layup with respect to the half-height of the profile. This was necessary to avoid any thermal distortion after extraction of the beam from the mold. The distance of the reinforcing rods from the external surface of the beam is 6 mm. Although this layup does not give the highest stiffening efficiency, the distance of 6 mm was chosen because it represents a minimum value necessary to obtain full rod wetting even in case of rod misalignment. A picture of the cross section of the reinforced beam obtained by the developed process is reported in Figure 6.

8.2 MATERIALS AND METHODS

The polymer material used in the present work is a recycled plastic coming from solid urban waste. In the framework of the Prowaste project, Inserplasa SL (Industria Sevillana de Reciclados Plasticos) was in charge of the production of RPL.

The RPL is obtained by manual sorting of plastics from solid urban waste. After removal of PET, HDPE and PP bottles, and of films of PP and PE bigger than about 20 × 30 cm, all the residues, mainly consisting of films of small dimensions, are collected as a mixed plastic. Such material mainly contains flexible and rigid PE and PP, but also small percentages of PET, escaped from the sorting stage. DSC analysis, reported in Figure 7, confirms that the material is mainly composed of LDPE, which melts in the range between 100 and 130 °C. Significant amounts of PP are also highlighted by the melting peak around 160 °C, as well as small traces of PET, which melts around 250 °C. Before extrusion, the material is simply milled to a size of about 8mm, and then pelletized at around 100 °C. No washing stage is foreseen.

Different types of pultruded glass rods have been used:

- Pultruded rods for optical applications, characterized by a thermoset matrix and a very smooth surface, were purchased from NEPTCO. The mechanical and physical properties of the NEPTCO rods are reported in Table 2. Pultruded rods of 3 and 4 mm in diameter were used. In order to obtain a rough surface, the 3 mm rods were roughened by sandblasting. These rods are referred to as NEPTCO_s. The micrographs obtained by optical microscopy of the as received 3 mm rods and sandblasted 3 mm rods are reported in Figure 8a.
- Pultruded rods for civil engineering applications, characterized by a thermoset matrix and a rough surface, were kindly supplied by POLYSTAL Composites. The mechanical and physical properties of the POLYSTAL rods are reported in Table 2. A microscope image of POLYSTAL fibers is reported in Figure 8b. As it can be observed, the rods are characterized by the presence of fibers aligned in the longitudinal direction, as well as of fibers aligned tangentially. These fibers contribute to the increase of surface roughness, and therefore are expected to increase the adhesion strength.
- Pultruded rods based on a thermoplastic matrix (polypropylene, PP) were supplied by JONAM composites. The mechanical and physical properties of the POLYSTAL rods are reported in Table 2. The micrograph of the pultruded JONAM rods are reported in Figure 8c, showing that the rods are actually made of a glass fibers core, surrounded by a PP matrix. On the other hand, a higher magnification image, reported in Figure 8d, clearly shows that each bundle is actually surrounded by the matrix. Nevertheless, each bundle is completely dry, and no matrix is present inside. Therefore it is possible to conclude that macro-impregnation occurs, but no micro-impregnation [14].

FIGURE 6: Picture of the cross section of the reinforced beam.

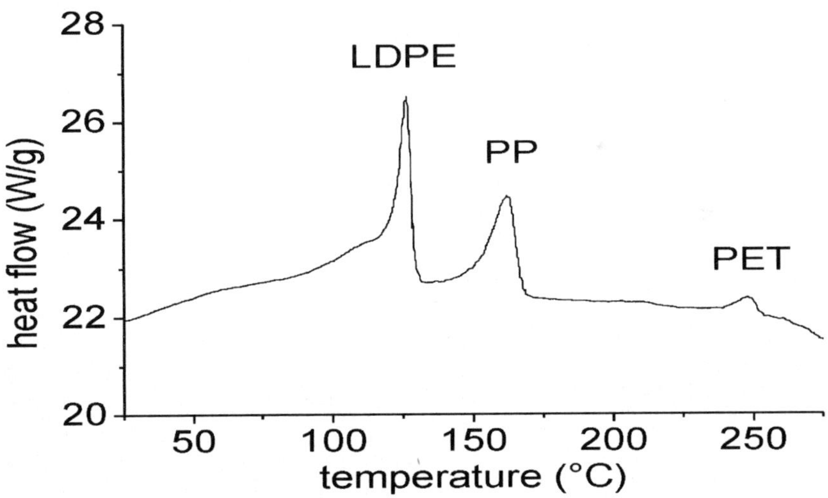

FIGURE 7: DSC analysis of RPL.

- The pultruded carbon reinforced rods from BASF series Mbar Joint are carbon fiber/epoxy composites with a tensile modulus of 70 GPa.

TABLE 2: Physical and mechanical properties of pultruded rods.

Supplier	Reinforcing Fibers	Matrix	Diameter (mm)	Tensile Modulus (GPa)
NEPTCO	glass (85% wt)	thermoset	3, 4	50
POLYSTAL	glass (85% wt)	thermoset	3	50
JONAM	glass (51% wt)	thermoplastic	6	7
BASF	carbon	thermoset	6	70

Samples for rod pullout tests were obtained by a double stage compression molding process. At first, 9 × 60 × 200 mm samples of RPL were obtained by compression molding under 200 bar and a plate temperature of 30 °C, after preheating the material at 170 °C. Then, the RPL plate was divided in two parts. The pultruded rods were enclosed between the two RPL plates and compression molded at 200 bar and a plate temperature of 30 °C, after preheating of the materials at 170 °C.

Pull-out tests were carried out according to ASTM D1871-98 standard, using a Lloyd LR5K dynamometer. Rectangular specimens 30 mm × 9 mm × 40 mm were cut from the compression molded samples. Each specimen has a single reinforcing rod, which protrudes 30 mm from the cross section area of the plastic mass. The crosshead speed for the pull-out tests was 50 mm/min. Pull-out tests were used to determine the adhesion strength (τs) between RPL and reinforcing rod, as:

$$\tau_S = \frac{F_{max}}{2\pi R L_0} \quad (4)$$

where F_{max} is the maximum load during the test, R is the rod diameter and L_0 is the contact length between rod and polymer mass. For comparison purposes, pull-out tests were also performed on beams reinforced with pultruded carbon reinforced rods and steel rods.

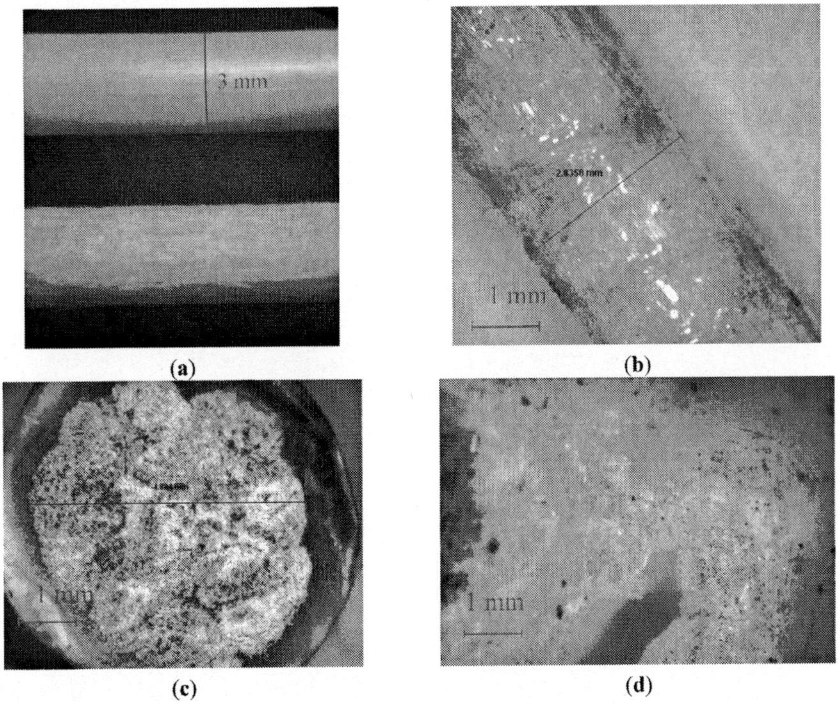

FIGURE 8: Micrographs of pultruded rods. (a) NEPTCO rods before (upper side) and after (bottom side) sandblasting; (b) POLYSTAL rods; (c) JONAM rods (cross-section surface); (d) JONAM rods after compression molding.

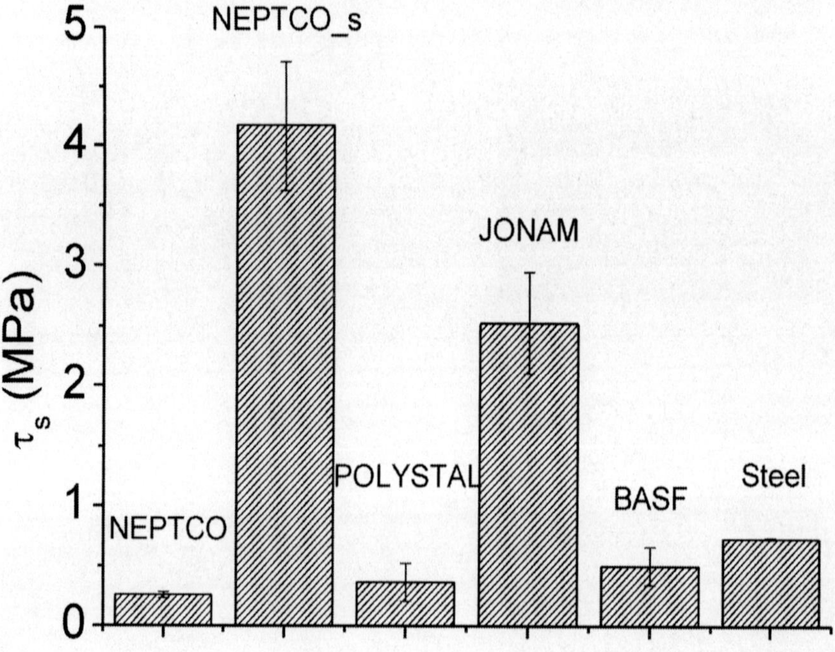

FIGURE 9: .Adhesion strength for different rods.

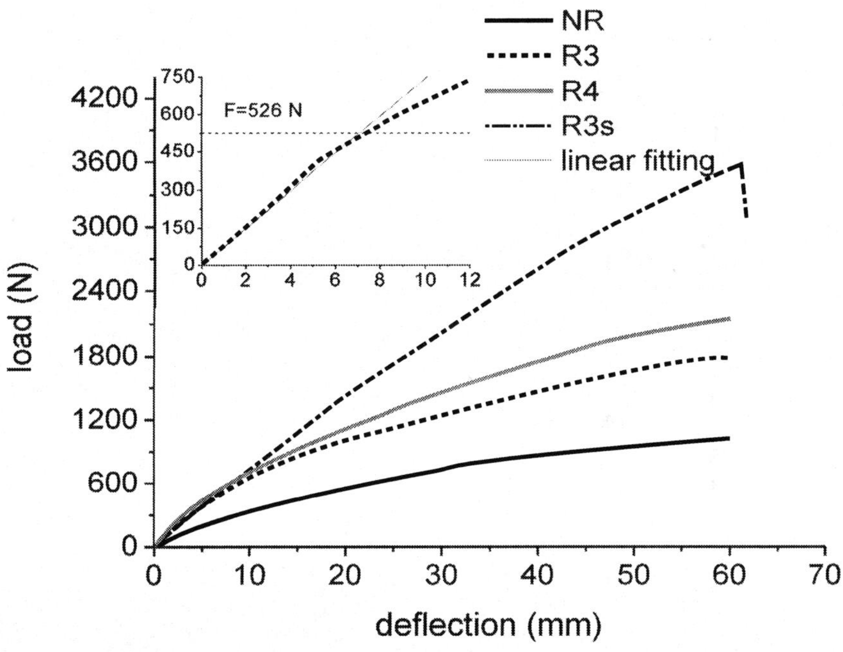

FIGURE 10: Load-deflection curves for plastic lumber beams.

Based on the results obtained from pull-out tests performed on the rods, thermosetting matrix based pultruded rods from NEPTCO were used for reinforcement of RPL. Four different prototypes were built. Beam NR was obtained by intrusion of the RPL without any reinforcement, beams R3 and R4 contained four rods of 3 mm and 4 mm in diameter, respectively, and beam R3s contained four 3 mm diameter sandblasted rods.

The specimens for flexural tests were obtained by cutting the beams along the length, to obtain samples with dimensions 40 mm × 120 mm × 800 mm. The span distance was 700 mm. For static flexural characterization, the sample was loaded in their middle section, with a crosshead speed of 10 mm/min, in accordance to ASTM D6109-97 standard. Cyclic flexural tests were performed between limits 0 to 1000 N, using a crosshead speed of 10 mm/min in both loading and unloading cycles. For flexural creep, test samples were placed in a oven and held at 50 °C for 1 h prior to testing. Then, the samples were loaded with a 240 N weight, and the deflection was recorded by means of a linear variable displacement transducer (LVDT) connected to a proprietary software, in accordance to ASTM D2990.

8.3 RESULTS AND DISCUSSION

The results of the pull-out tests performed on the different rods are shown in Figure 9. The adhesion strength for the rods based on thermoset matrix, i.e., NEPTCO and POLYSTAL, is quite poor, which is due to the fact that during processing, the matrix of the rod remains in the solid state, which prevents a good adhesion to the RPL. Most likely, the smooth surface of the rods (even those characterized by the presence of winded fibers, POLYSTAL) does not allow any mechanical gripping of the surface of the rod to RPL. On the other hand, the use of carbon and steel rods characterized by a rough surface, which are specifically designed for the building sector where adhesion to concrete is a key issue, allows for improvement of the adhesion to RPL. For the same reason, NEPCTO_s rods show an adhesion strength which is about 1 order of magnitude higher than that of the corresponding rods before sandblasting, NEPTCO. The difference is only a consequence of the different surface roughness of the materials. Instead, for JONAM rods, the improved adhesion can be attributed to the melting of the matrix during processing. The melting allows for some inter-diffusion at the interface between RPL and rod, which in turn involves an increase of the adhesion. In view of the results obtained, and also accounting for the cost considerations reported in Table 1, NEPTCO rods were used for the production of selectively reinforced RPL on the industrial plant.

The results from static flexural tests are reported in Figure 10. The behavior of the sample NR is highly non linear. However, from the slope of the curve at the origin, a modulus of 395 MPa can be calculated. When pultruded rods are incorporated in the plastic lumber beam, the flexural stiffness of the beam significantly increases.

By coupling Equations (1) and (2), the stiffness of the beams reinforced by the 3 mm and 4 mm diameter rods was estimated to be 74 and 100 N/mm, respectively. The linear prediction according to the estimated stiffness are reported in the inset of Figure 10 for beam R3, showing a very good agreement with the experimental curves. Figure 10 also shows a slope change for samples R3 and R4 at a load level of about 500 N. This discontinuity can be attributed to debonding of the rods as the shear stress exceeds the adhesion stress between matrix and rod. This occurs when:

$$\tau_{rz|R} = \tau_s \quad (5)$$

which, according to Equation (3), occurs when:

$$F_{debonding} = K\frac{\tau_S L^3}{12E_{ROD}cR} \quad (6)$$

The value of F causing debonding, evaluated according to Equation (6), is 526 N, which is a value in a very close agreement to the experimental value, as evidenced in the inset of Figure 10.

For sample R3s, the properties at low stress levels are roughly the same of sample R3, indicating that the improvement of rod adhesion does not involve an improvement of the stiffness of the beam. In both cases, at low levels of deformation, the stress is transferred between the two phases by elastic shear, and there is perfect adhesion between the two phases. On the other hand, no discontinuity is observed in Figure 10 for the load-displacement curve of sample R3s, indicating the absence of debonding or slip effects at the interface between rod and RPL. Beam failure is observed under an applied load of about 3600 N. In correspondence of 3600 N load, Equation (3) allows to estimate a shear stress at the rod-RPL interface of 1.77 MPa, which is about 1/3 of the experimental value of adhesion strength reported in Figure 9 for sandblasted rods. This confirms that beam failure is not due to debonding at the rod/RPL interface. Indeed, the normal stress on the rod can be estimated to be [12]:

$$\sigma_z = 12F\frac{cE_{ROD}}{KL^2} \tag{7}$$

yielding a value of about 830 MPa, which is quite close to the tensile strength of the rods (about 1.4 GPa, as reported in the technical data sheet). This observation indicates that, for sample R3s, beam failure is due to rod tensile failure. For the samples R3 and R4, failure by flexural break did not occur even at displacements as high as 60 mm, as shown in Figure 10. This is the result of the very poor stress transfer between rod and polymer.

The results from flexural characterization suggest that failure of the reinforced RPL can occur due to two different phenomena:

1. Debonding at the RPL-rod interface. This failure mode is very similar to yielding of ductile materials, since it involves a permanent plastic deformation, but does not involve a sudden decrease of the load bearing capacity. Debonding occurs when the force equals the value reported in Equation (7).
2. Tensile failure of the rods. This failure mode is very similar to the rupture of a brittle material, since it involves a sudden decrease of the load bearing capacity of the beam. Tensile failure of the rods occurs when the normal stress on the rods equals their tensile

strength, σR,ROD, and the force attains a value given by inversion of Equation (7):

$$F_{ROD} = \frac{\sigma_{R,ROD} L^2 K}{12 E_f c} \tag{8}$$

The lower value of the forces calculated according to Equations (6) and (8) determines the mode of failure of the reinforced beam. The results are reported in Table 3 for rods with 3 mm diameter in a beam 120 mm × 40 mm × 700 mm. As it can be observed, the NEPTCO rod-reinforced RPL fails due to debonding at the interface, and the same is likely to occur for POLYSTAL and JONAM rods. Despite this, the JONAM rods-reinforced RPL is likely to fail at much higher values of the applied load, as reported in Table 3, due to the much higher adhesion strength. In contrast, the JONAM rods-reinforced RPL is characterized by lower values of the stiffness, due to the lower modulus of JONAM rods compared to NEPTCO ones.

TABLE 3: Mode of failure of pultruded rod reinforced RPL.

Flexural properties of reinforced beams	NEPTCOϕ3	NEPTCO Sandblastedϕ3	POLYSTALϕ3	JONAMϕ3
E_R (GPa)	50	50	50	15
τ_s (MPa)	0.26	4.17	0.36	2.5
$F_{debonding}$ (N)	526	8433	728	10,780
F_{rod} (N)	6067	6067	6067	12,834
Mechanism of failure	Debonding @ 526 N	Fiber tension @ 8433 N	Debonding @ 728 N	Debonding @ 10,780 N
Flexural stiffness (N/mm)	74	74	74	47

The results reported in Table 3 suggest that JONAM rods should be used when high load bearing capability is the most important design parameter, whereas NEPTCO rods should be chosen when a high stiffness is more relevant. The use of sandblasted NEPTCO rods allows for the obtainment of a very good compromise between high stiffness and high load bearing capacity. In this view, the choice between sandblasted and not sandblasted rods should only be made based on the economics of sandblasting, which is behind the scope of this work.

The results reported in Table 3, calculated according to Equation (7) and (8), do not account for the length of the beam. In fact, the results are relative to 700 mm long beams, which were used for mechanical characterization, though in most cases RPL beams can be as long as 2000 mm. In such cases, combining Equation (7) and (8) yields:

$$\frac{F_{rod}}{F_{debonding}} = \frac{\sigma_{R,ROD}}{\tau_S} \frac{R}{L} \tag{9}$$

which gives a very useful tool for failure prediction. If the ratio $F_{rod}/F_{deboning}$ is higher than 1, failure occurs due to debonding, whereas rod tensile failure can occur when the ratio is lower than 1. For example, for JONAM rods, at a beam length of 700 mm, $F_{rod}/F_{deboning} = 1.2$ indicates that failure occurs due to debonding, as reported in Table 3. On the other hand, for a beam length of 2000 m, $F_{rod}/F_{deboning} = 0.42$ indicates that failure is likely to occur due to rod tension.

For a better understanding of the concept of debonding, and to highlight its analogy to the yielding of a ductile material, cycling loading tests were performed and the results are reported in Figure 11. The RPL reinforced with NEPTCO rods shows some important features. As also observed for the loading tests, a change of the slope in the curve at 530 N indicates the presence of debonding phenomena. After loading up to 1000 N, a significant permanent displacement (about 9.20 mm) is very similar to a plastic deformation for a ductile material. On the other hand, when RPL is reinforced with sandblasted NEPTCO rods, no change of the slope is observed during the loading stage. Consequently, the residual displacement after the unloading step is reduced to 0.9 mm. This is equivalent to a perfectly elastic behavior of the beam R3s between 0 and 1000 N.

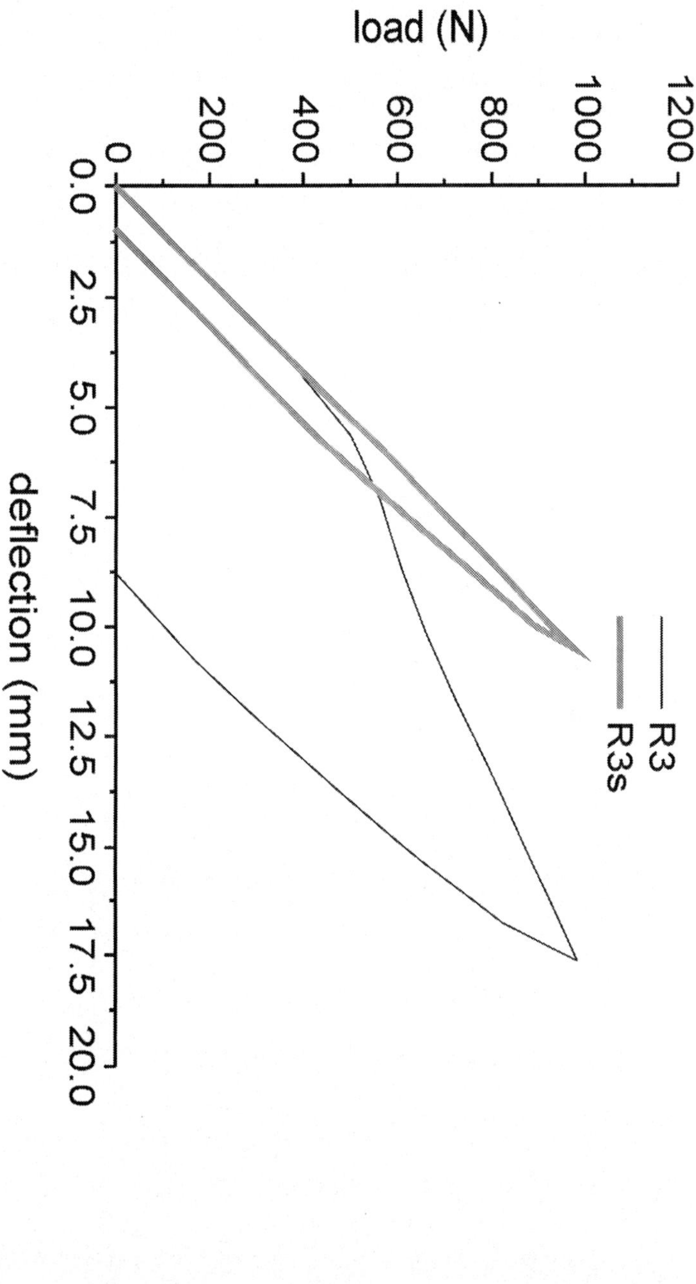

FIGURE 11: Cyclic loading tests for plastic lumber beams.

Finally, it is possible to compare the present technology with other approaches which can be used for RPL stiffening:

1. Addition of glass spheres, holding the same geometry of the beam.
2. Increase of the thickness of the beam, holding the same material (unreinforced RPL).

For each layout of the reinforcing rods, it is possible to calculate the stiffness ratio, as the ratio between the stiffness of reinforced RPL, KR, and the stiffness of unreinforced RPL, KNR:

$$SR = \frac{K_R}{K_{NR}} \tag{10}$$

Addition of the glass spheres yields to an isotropic and homogeneous material, in which the reinforcement is also disposed in zones subjected to very low stresses. For each value of the stiffness ratio, the amount of glass spheres to be added can be obtained by inversion of the Halpin-Tsai equation [15]:

$$SR = \frac{1 + 2\dfrac{\dfrac{E_f}{E_{RPL}} - 1}{\dfrac{E_f}{E_{RPL}} + 2} v_f}{1 - 2\dfrac{\dfrac{E_f}{E_{RPL}} - 1}{\dfrac{E_f}{E_{RPL}} + 2} v_f} \tag{11}$$

in which the elastic modulus of glass, E_F, is 72 GPa, that of RPL, E_{RPL}, is 395 MPa, and the aspect ratio of the reinforcing particles is assumed to be the unity. Assuming a plastic lumber density of 920 kg/m^3 and a glass density of 2540 kg/m³, it is finally possible to calculate the corresponding weight of the beam.

The equivalent increase of the thickness of the beam can be calculated starting by the modulus of the material:

$$SR = \frac{h_{NR}^3}{h_R^3} \tag{12}$$

where: h_R is thickness of the pultruded rod reinforced beam, and h_{NR} the thickness of the unreinforced beam characterized by the same stiffness. The equivalent weight increase can be estimated considering the new geometry of the beam. The results are reported in Figure 12 for the three different methods. As it can be observed, the Prowaste concept allows to optimize the stiffening efficiency, reducing the weight increase to less than 4% for a layup in which 10 rods are disposed symmetrically, with a stiffening ratio SR = 3.7. Alternative methods, which introduce an uniform and homogeneous reinforcement, are less effective in increasing the stiffness, and involve a significant increase of the weight of the component.

Finally, the creep curves for reinforced beams are reported in Figure 13. Compared to NR beam, the addition of the NEPTCO rods involves a reduction of the deflection of the beam by a factor of about 1.5. On the other hand, sandblasting of the rods has a dramatic effect, since it involves a reduction of the deflection measured after 25 h from 18.2 mm to 1.32 mm. This indicates that, in the case of R3 beam, slip at the rod/RPL interface plays a major role.

Finally, a picture of a bench produced by reinforced RPL is reported in Figure 14. In the framework of the Prowaste project, CETMA consortium (Brindisi, Italy) was in charge of the design and assembling of the bench. The introduction of the reinforcing rods allowed for the production of a bench characterized by a very long span (in this case almost 2000 mm) whereas for standard RPL span no longer than 800–900 mm are suggested. The bench was designed with 6 profiles fixed on 2 basements with a comb interlocking system, allowing easy assembly and different ways to seat.

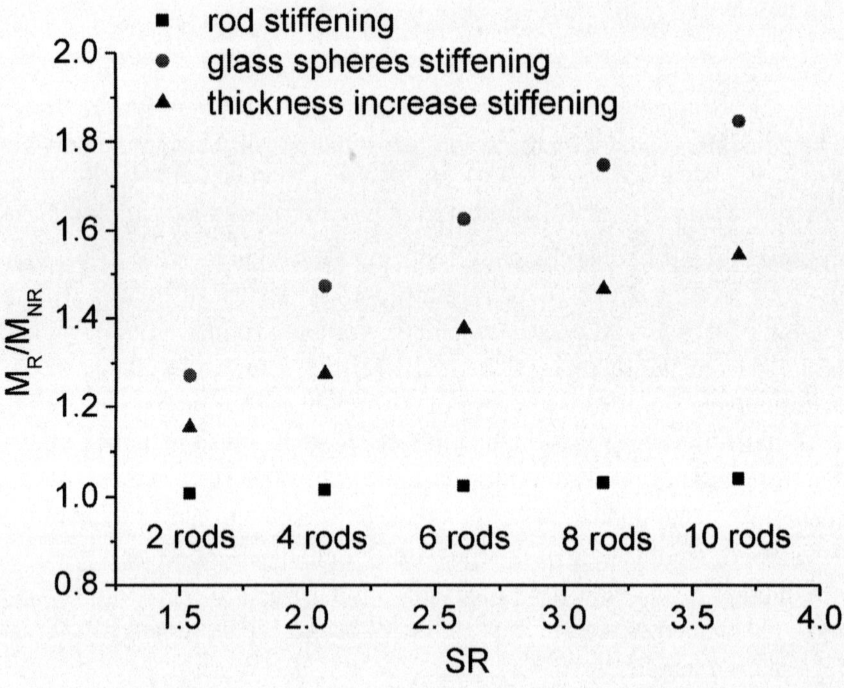

FIGURE 12: Comparison of the different stiffening approaches

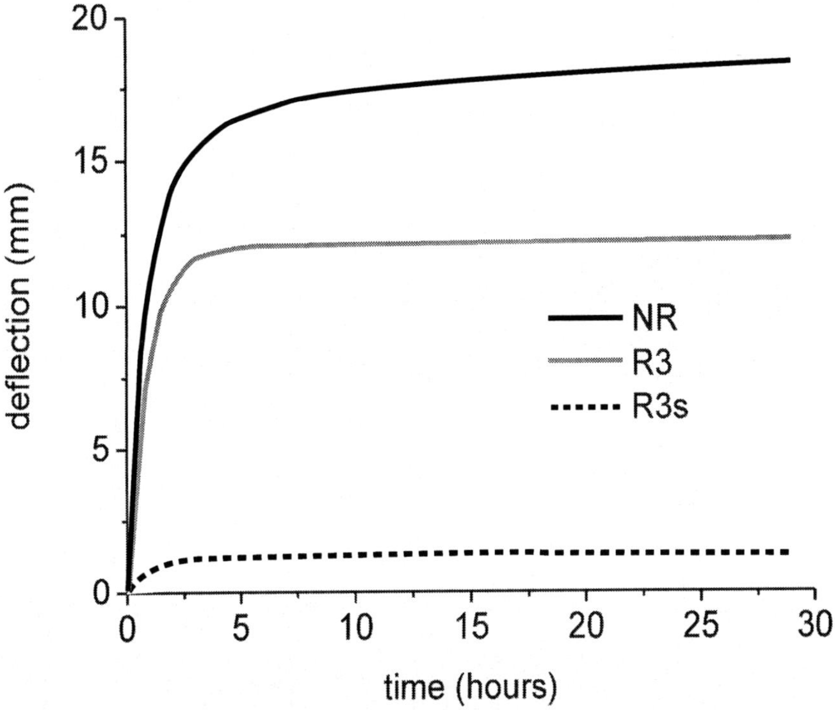

FIGURE 13: Creep curves for plastic lumber beams.

FIGURE 14: Picture of the bench produced with Prowaste profiles.

8.4 CONCLUSIONS

The Prowaste project introduced a new design and process adaptation for the production of selectively reinforced recycled plastic lumber beams. The basic concept is based on the use of high aspect ratio reinforcing elements, placed in specific points of the cross section of the beams. Based on a primary evaluation of different classes of materials, pultruded glass rods are the best candidates for the proposed application. The process adaptation is based on the use of two metallic frames, the first one fixed on the front surface of the mold, and the second one capable to slide inside the mold with melt front during mold filling. The use of the two tools allows for disposal of the reinforcing rods parallel to the surface of RPL beam, at a specific distance from the half height of the beam cross section. Pull-out tests showed that thermosetting matrix pultruded rods are characterized by a very low adhesion strength to RPL. On the other hand, sandblasting can improve by one order of magnitude the adhesion strength. Instead, thermoplastic matrix rods, though being characterized by a low stiffness compared to thermosetting matrix rods, show a very good adhesion to RPL.

Flexural tests showed that the incorporation of pultruded glass rods in the RPL beams significantly improves the flexural stiffness. For thermosetting matrix rods having smooth surface, the poor interfacial adhesion is responsible of debonding at the polymer/rod interface when the interfacial stress overcomes adhesion strength. Debonding can be prevented by roughening the surface of the pultruded rod. This suggests that, based on the debonding behavior of reinforced RPL, the beam can behave either as a ductile or as a brittle material. In fact, debonding causes a significant non linearity, with the presence of significant plastic deformations, which is mechanically similar to yielding of a ductile material. In such circumstances, failure of the beam is not catastrophic, which is also very similar to ductile materials. On the other hand, a very high adhesion strength causes failure to occur due to tensile rupture of the rod, which in turn involves a sudden decrease of the load bearing capability. Similarly, a very high adhesion prevents debonding, and therefore plastic deformations. In such cases, the beam behaves like a brittle material.

A comparison with other stiffening approaches show that the Prowaste concept allows to minimize the weight increase, which is a key issue for RPL.

Finally, the Prowaste concept allowed for the production of a bench with an innovative design, in which, due to the high stiffness of the beam, the span was increased up to 2000 mm, which is a value much higher than that commonly used for RPL beams.

REFERENCES

1. Stein, R. Miscibility in polymer recycling. In Emerging Technologies in Plastics Recycling; Andrews, G.D., Subramanian, P.M., Eds.; ACS Series: Washington, DC, USA, 1992.
2. Cao, L.; Ramer, R.M.; Beatty, C.L. Evaluation of mechanical properties of recycled commingled post materials. In Emerging Technologies in Plastics Recycling; Andrews, G.D., Subramanian, P.M., Eds.; ACS Series: Washington, DC, USA, 1992.
3. Blatz, P.S. Properties of HDPE from postconsumer recycled container. In Emerging Technologies in Plastics Recycling; Andrews, G.D., Subramanian, P.M., Eds.; ACS Series: Washington, DC, USA, 1992.
4. Greco, A.; Frigione, M.E.; Maffezzoli, A. Rotational moulding of recycled HDPE. Polym. Recycl. 2001, 6, 23–33.
5. Greco, A.; Maffezzoli, A.; Manni, O. Development of polymeric foams from recycled polyethylene and recycled gypsum. Polym. Degrad. Stab. 2005, 90, 256–263.
6. La Mantia, F.P.; Perrone, C.; Bellio, E. Recycling of Plastic Materials; ChemTech Publishing: Toronto, ON, Canada, 1993.
7. Leidner, J. Plastic Waste; Marcel Dekker: New York, NY, USA, 1991.
8. Brandrup, J.; Bittner, M.; Michaeli, W. Recycling and Recovery of Plastics; Hanser Publishers: Munich, Germany, 1992.
9. March, F.A.; Taylor, R.B.; Menge, J.H.; Gould, R.J.; Pontiff, T.H. Elongated Structural Member and Method and Apparatus for Making Same. U.S. Patent 5,650,224, 22 July 1997.
10. Balazek, D.T.; Griffiths, T.J.; Pearson, D.E. Pultrusion/Extrusion Method. U.S. Patent 4,938,823, 3 July 1990.
11. Borzakian, V. Plastic piling. U.S. Patent 5,051,285, 24 September 1991.
12. Pio, C.; Greco, A.; Maffezzoli, A.; Marseglia, A. Efficient utilization of plastic waste through product design and process adaptation: A case study on stiffness enhancement of beams produced from plastic lumber. Adv. Polym. Tech. 2008, 27, 133–142.
13. Greco, A.; Romano, G.; Maffezzoli, A. Selective reinforcement of LLDPE components produced by rotational molding with thermoplastic matrix pultruded profiles. Compos. Part B Eng. 2014, 56, 157–162.
14. Gennaro, R.; Greco, A.; Maffezzoli, A. Micro- and macro-impregnation of fabrics using thermoplastic matrices. J. Thermoplast. Compos. 2011, 26, 527–543.
15. Halpin, J.C.; Kardos, J.L. The Halpin-Tsai Equations: A Review. Polym. Eng. Sci. 1976, 16, 344–352.

CHAPTER 9

Recycling Glass Cullet from Waste CRTs for the Production of High Strength Mortars

STEFANO MASCHIO, GABRIELE TONELLO, AND ERIKA FURLANI

9.1 INTRODUCTION

CRT glass waste includes that from TVs, PC monitors and other monitors used in special applications, and waste from the original assembly process. Waste glass from PC and TV monitors will begin to decline as a direct consequence of the emerging flat screen display technology; nevertheless, it seems reasonable to assume that an amount of CRTs from all sources is likely to continue to enter into the waste stream in the coming years. Studies have shown that when CRTs are disposed of in landfill sites, leaching processes from the crushed glass cullet may contaminate ground water. This is a major driving force for CRT recycling. Moreover, it must be pointed out that CRT waste does not contain only glass but also other materials which concur with the CRTs assembly, such as ferrous and nonferrous metals and plastics. The Waste Electrical and Electronic Equipment (WEEE) directive sets strict regulations for recycling or recovery when materials derive from equipment containing CRTs. Such norms must obviously be coupled with those reported by the European Waste Catalogue which classifies CRTs as hazardous waste and makes landfill disposal of CRT materials costly. The great amount of CRT waste produced all over the world implies that its recycling is presently necessary not only due to the rising cost of landfill disposal, which is reflected on the cost of new CRTs produced, but also as a consequence of the "zero-waste" objective which must be the final goal of all future human activities. Mixed CRT glass (funnel, neck and screen glass) contains high amounts of PbO, BaO, and SrO; it follows that cullet with this composition is unsuitable for recycling in applications where metal oxides could leach into food products or ground water. A possible partial recovery of waste CRT glass from the assembly process could follow the path of the manufacture of new CRTs, even if the high cost of separating, sorting, and processing the glass to meet the standards required by glass manufacturers strongly limits this option [1]. Other methods for waste CRT glass recycling are the copper-lead smelter where it acts as a substitute for sand in the smelting process [2] or as an additive raw material in the ceramic industry for the production of tiles or other monolithic ceramic materials [3, 4]. The above options represent, however, a small number of opportunities for waste CRT glass recovery, and additional proposals are necessary in order to maximise its reuse in the production of other types of materials.

In order to propose a new option for waste CRT glass recycling, as independently as possible from glass composition, in the present research the production of stable mortars was carried out using cement, ground-waste-mixed CRT glass (funnel, neck, and screen glass), natural aggregate, and water; the addition of superplasticizer was also investigated. Mortars are expected to take advantage of ground glass; it is known, in fact, that the addition of silica fume as a component material in the production of mortars or concretes can lead to the preparation of products, namely, reaction powders mortars or concretes (RPM, RPC), with low water absorption, high mechanical properties, and modified shrinkage [5, 6]. In a parallel approach, the addition of generic waste glass to mortars or concretes has been widely investigated [7, 8]. It has been demonstrated, for example, that the use of waste glass from crushed containers as concrete aggregate may develop pozzolanic activity which is affected by glass finenesses and chemical composition [9, 10]. Such properties may also affect workability [11], strength, and durability. More in particular a high content of alkalis can cause alkali silica reaction (ASR) and expansion [12, 13]. Conversely, the use of waste E-glass from electronic scraps (low-alkali content) improves compressive strength and sulphate resistance and reduces chloride-ion penetration with no adverse alkali silica reaction-expansion effects [14]. The addition of the specific CRT glass waste has been, on the other hand, recently studied [15–17]; however additional detailed studies are worthy of interest.

In the present research, mortars were produced using a fixed cement/aggregate (c/a) ratio (1/3), as it has been often proposed in the literature [18, 19], whereas milled CRT mixed glass was added in different proportions as well as were different contents of superplasticizer. The goal of the present research is to demonstrate that by selecting a proper amount of milled CRT waste glass coupled with an optimal quantity of superplasticizer, it is possible to produce mortars with high compressive strength, low water absorption, and therefore long durability coupled with low elution release of hazardous elements.

9.2 EXPERIMENTALS

9.2.1 MATERIALS

The starting materials used were: a type I ordinary portland cement (OPC) and a natural aggregate with maximum particle dimension of 4.76 mm, Blaine fineness of 3480 $cm^2 g^{-1}$ (EN 933-2), density of 2.46 $g cm^{-3}$, and water absorption of 0.37% (the measurement was carried out following the ASTM C127 and C128 norms) which were mixed with different proportions of mixed CRT waste glass. The as-received CRT glass cullet was first transformed into a powder by milling and then sieved by a 500 μm sieve. Only the part of powdered glass with size smaller than 500 μm was used for the present research; particles of larger dimensions were remilled. Glenium 51 (BASF) was also used as a superplasticizer in the preparation of some specimens. The required amount of water was obviously added to each starting blend. The chemical analysis of cement, natural aggregate, and CRT glass, determined by a Spectro Mass 2000 ICP mass spectrometer, is reported, in terms of oxides, in Table 1 which also displays loss on ignition (LOI), obtained after thermal treatment at 1000°C for 2h, density, water absorption, and Blaine fineness.

TABLE 1: Composition (oxide wt%), organic carbon, specific gravity, LOI and fineness modulus of milled CRT mixed glass, cement and aggregate; "undetermined" indicates the cumulative quantity of all oxides determined in quantity lower than 0.1 wt%.

Component	CRT glass	Cement	Aggregate
SiO_2	59.74	21.40	1.98
Al_2O_3	2.67	1.48	1.72
CaO	2.06	61.02	46.73
MgO	1.15	1.16	19.83
Na_2O	6.81	0.26	1.43
K_2O	6.15	0.53	0.74
Fe_2O_3	0.11	0.35	2.1
TiO_2	0.17	<0.1	<0.1
CuO	0.21	<0.1	<0.1
BaO	6.13	<0.1	<0.1
SrO	4.77	<0.1	<0.1
Sb_2O_3	0.28	<0.1	<0.1
ZrO_2	0.39	<0.1	<0.1
PbO	8.33	<0.1	<0.1
SO_4^-	0.49	2.92	0.13
C (organic)	<0.1	1.20	0.57
Undetermined	0.93	1.38	1.92
Density ($g\,cm^{-3}$)	2.95	3.03	2.46
Water abs. (%)	0.20	—	0.37
LOI (%)	0.82	13.14	23.55
Fineness modulus	0.72	—	3.48

It is observed that CRT glass contains, together with an expected major quantity of SiO_2, also great fractions of PbO, Na_2O, K_2O, BaO, SrO, and moderate quantities of Al_2O_3, CaO, and MgO; other compounds as well as organic carbon are present in limited amounts so that also LOI is limited as is the water absorption. Aggregate mainly contains calcium and magnesium oxide accompanied by small fractions of silica, iron oxide, and alumina; organic carbon, density, and LOI are in line with the literature data [5, 20]. The OPC conforms to European Standards EN-197/1. Data reported in Table 1 are confirmed by the XRD analysis of the starting materials (not reported in the present paper) which revealed the presence of alite (84%) and belite (16%) in cement, whereas dolomite (65%), calcium carbonate (27%), and free quartz (8%) were identified in the aggregate; numbers must be read with caution since XRD analysis does not provide accurate quantitative analysis of the tested materials, but they only supply an approximate magnitude order of their crystallographic composition. It can be also observed that CRT glass has a density of 2.95, OPC 3.03, and aggregate 2.46 $g\,cm^{-3}$. It is worth to point out the low Blaine fineness of the milled and sieved CRT glass (0.72), thus confirming the presence of a large fraction of particles with size below 75 μm and therefore in agreement with the characteristics suggested by other authors [14] when the production of high-performance materials containing milled waste glass is required.

9.2.2 METHODS

9.2.2.1 X-RAY DIFFRACTION INVESTIGATION (XRD)

The crystalline phases of starting components as well as those of the hydrated materials were investigated by X-ray diffraction (XRD). XRD patterns were recorded on a Philips X'Pert diffractometer operating at 40 kV and 40 mA using Ni-filtered Cu-K_α radiation. Spectra were collected using a step size of 0.02° and a counting time of 40 s per angular abscissa in the range of 15–55°. Philips X'Pert High Score software was used for phase identification and semiquantitative analysis (RIR method).

9.2.2.2 PARTICLE SIZE DISTRIBUTION (PSD) MEASUREMENTS

The particle size distribution (PSD) of the fine fraction of the aggregate (<500 μm), cement, and powdered mixed CRT glass were determined by a Horiba LA950 laser scattering PSD analyser; analyses was made in water after a 3 min sonication; PSD curves are represented with logarithmic abscissa. In order to access the PSD of the aggregate's fine fraction, the total as received product was sieved (500 μm) and fines were separated from coarse particles; the fines represent 25% of the total aggregate.

9.2.2.3 MATERIALS COMPOSITION

The ratio between cement and aggregate quantity (natural aggregate plus glass cullet) was set at 1/3 as this is a frequently used ratio. Some reference glass free compositions, hereafter called R, containing cement, aggregate, superplasticizer, and an optimized amount of water were also prepared as blank samples in order to compare the mechanical behaviour, after hydration, of the materials produced, bearing in mind that the focus of the present research regards the production of materials obtained by replacing part of the natural aggregate with an equivalent mass of 5, 10, and 20 wt% of milled and sieved glass powder from mixed CRT glass. Samples with symbolic names, corresponding aggregate composition, s/c, and water/cement (w/c) ratios are reported in Table 2.

9.2.2.4 MATERIALS PREPARATION

For the mixture preparation and w/c optimization, a 5 L Hobart planetary conforming to ASTM C305 standards was used. The optimized amount of water was determined by the ASTM C1437 slump test performed on the reference blend R. The paste is said to have the right workability if the cake width is 200 (±20) mm. The identified optimal w/c value of the reference blend (R) was 0.44; this same value was applied to all the superplasticizer free compositions. Blends containing superplasticizer required reduced amounts of water as displayed in Table 2. Pastes were then poured under vibration into moulds with dimensions of 100 × 100 × 100 mm, sealed with a plastic film to ensure mass curing and aged 24 h for a first hydration. Samples were then demoulded, sealed again with a plastic film, and cured again in the air for 24 h and then in water at room temperature for 3, 7, 28, 90, and 180 d. The ageing water was maintained at the constant temperature of 25°C (±3°C) and replaced with fresh water every 3 d of curing. After curing, before their characterisation, samples were dried with a cloth and aged in the atmosphere for 24 h. Specimens used for release evaluation were not aged in water but sealed with a plastic film for 7 d and then tested.

TABLE 2: Specimens symbolic names, corresponding aggregate composition, superplasticizer/cement (s/c) and water cement (w/c) ratios.

Sample	Natural aggregate (wt%)	CRT glass (wt%)	s/c (%)	w/c
R	100	0	0	0.44
R1	100	0	1	0.31
R2	100	0	2	0.27
V5	95	5	0	0.44
V51	95	5	1	0.31
V52	95	5	2	0.27
V10	90	10	0	0.44
V101	90	10	1	0.31
V102	90	10	2	0.27
V20	80	20	0	0.44
V201	80	20	1	0.31
V202	80	20	2	0.27

9.2.2.5 MATERIALS CHARACTERIZATION

Compression tests were performed after 3, 7, 28, 90, and 180 d in accordance with the ASTM C469 norm using Shimadzu AG10 apparatus; data were averaged over 5 measurements. Expansion was measured by a caliper after 28 d of curing in water. Fracture surfaces were examined by an

Assing EVO40 Scanning Electron Microscope (SEM) coupled with the Energy Dispersive X-ray Spectroscopy (EDXS). The ASTM C642 norm was used to test the water absorption of the samples after curing for the established number of days.

9.2.2.6 *LEACHING EVALUATION*

After ageing for 7 d, the R and V20 samples were submitted to an elution release test in water. V20 was selected in order to test a composition containing the highest amount of mixed CRT glass and displaying a high level of water absorption. For this measurement, the above specimens were aged in an autoclave, at 120°C, and 2 kPa for 1 h using 4 L of water. After boiling, samples were cooled down to room temperature (in water). The solutions (water plus several salts eluted from the samples weight ratio between mortar sample and solution is around 0.17) were submitted to the ICP tests. For comparison, materials aged in water for 28 d and used for testing the mechanical strength were submitted to the Pb release which was determined according to the US EPA 1311 TCLP test which establishes to crush the hydrated mortar (28 d), sieve the product through a 10 mm sieve, and extract a solution using glacial acetic acid.

9.3 RESULTS AND DISCUSSION

The starting materials have different particle size distribution: the OPC displays a monomodal distribution of particles with maximum concentration at 12 μm (see Figure 1); milled and sieved CRT glass has a broad PSD with a very large peak at around 180 μm, therefore showing the presence of about 60 vol% of coarse particles with size greater than 75 μm and around 40 vol% of fines with size below 75 μm; the fine fraction of the natural aggregate displays a bimodal distribution of particles with two peaks: one, low at 11 μm, the other, higher, at around 130 μm. It is known that the most reactive fraction of a component is its finest fraction, which also affects the w/c ratio as well as the alkali silica reaction in the presence of Na_2O and K_2O. In this contest, we have considered it important to show that the fine fraction of the milled waste glass falls in the range of sub-micronic sized particles and is overlapped to that of the smallest cement particles. Also natural aggregate contains small particles, the amount of alkalis is very low, and their size is always greater than 2 µm. It therefore appears reasonable that waste glass particles with size below 1 µm could easily interact with cement particles of the same dimension developing pozzolanic activity, limiting ASR, and improving long term properties of the resulting materials. Conversely, small sized particles also influence w/c ratio reducing, in this way, mortars workability. However, the absolute amount of glass fines is limited so that their influence on the fresh mortars workability as well as on the properties of the resulting hydrated materials is expected to be limited.

Figures 2(a), 2(b) and 2(c) show the trend, displayed as a function of curing time, of compressive strength (solid lines, left axis) and water absorption (dashed lines, right axis) of three sets of materials, respectively: (a) superplasticizer free, (b) with an s content of 1%, and (c) with an s amount of 2%. Error bars are, for clarity, not displayed due to their overlapping and the possible confusion on reading; however data scattering is, for each composition, maintained within the interval ±5% of the reported average value. Figure 2(a) shows that the average compressive strength of the reference R reaches 50 MPa after 3 d of curing, 66 after 7, 87 after 28, and 97 after 90 and rises to 100 after 180 d; the corresponding water absorption is 6.3% after 3 d of curing, 5.7 after 7, and 5.4 after 28 and lowers to 5.1 and 4.9 after 90 and 180 d, respectively. Such an abnormally high compressive strength is reasonably due to the high strength, low porosity (low water absorption), and chemical nature of the aggregate used in the present research [21]. Compressive strength of composition V10 raises from 45 MPa after 3 d of curing to 60 after 7, 83 after 28, 100 after 90, and 105 after 180 d; water absorption ranges from 7.1 (3) to 4.4% (180). It can be observed that compositions V5 and V20 show compressive strength curves below that of the reference material over the whole range of time, whereas that of composition V10 intercepts it between 28 and 90 d of curing, the final strength being higher than that of R; the addition of CRT glass improves its long-term strength which is little affected by the possible ASR.

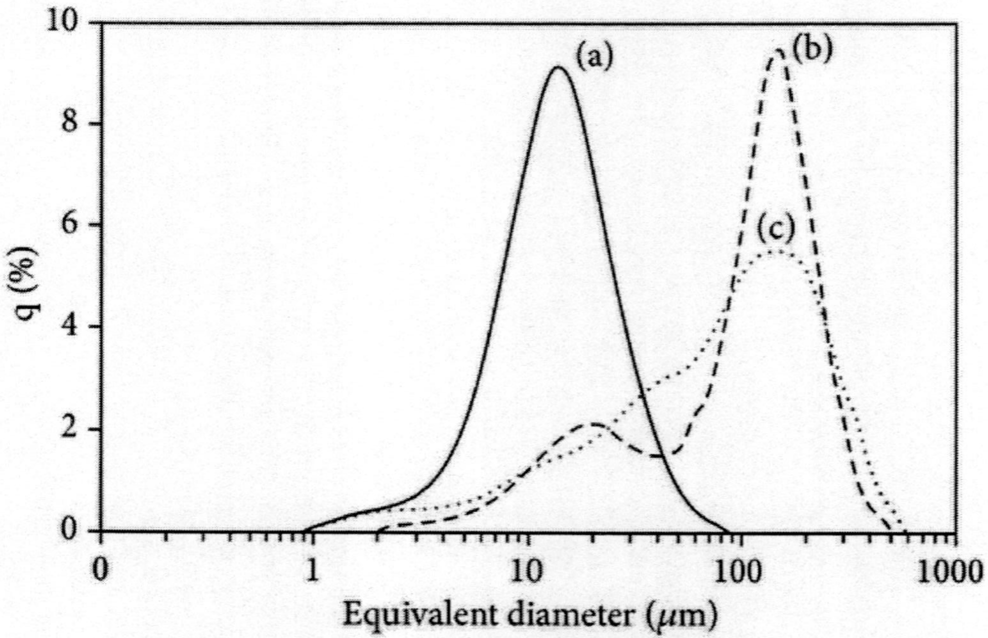

FIGURE 1: PSD curves of the OPC (a), the fine fraction of aggregate (b), and that of the milled and sieved mixed CRT glass used for samples preparation (c).

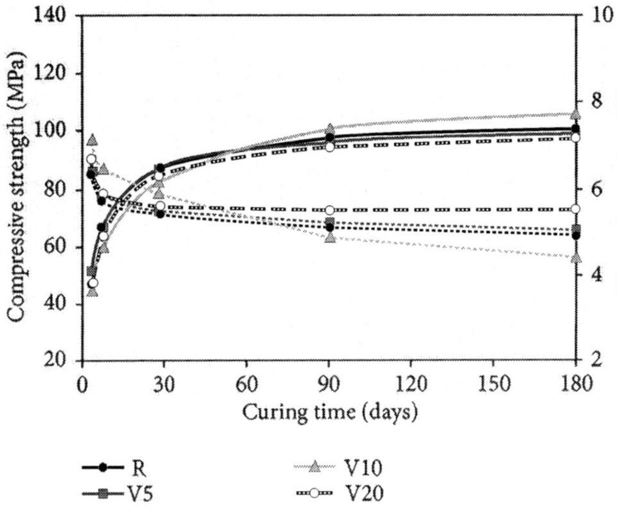

A

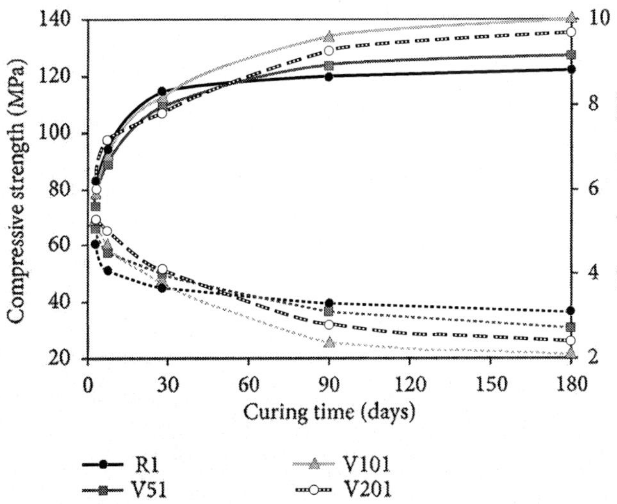

B

FIGURE 2: Trend, displayed as a function of curing time, of compressive strength (solid lines, left axis) and water absorption (dashed lines, right axis) of three sets of materials: (a) superplasticizer free; (b) with a superplasticizer content of 1%; (c) with superplasticizer content of 2%. Error bars are, for clarity, not displayed due to their overlapping and the possible confusion on reading.

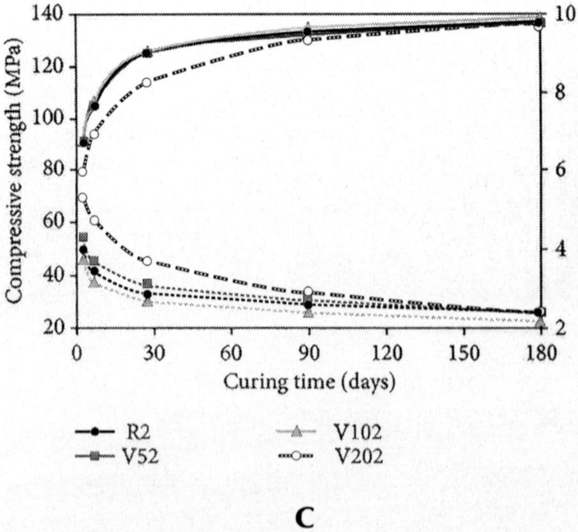

FIGURE 2: *Cont.*

In fact, Chen et al. demonstrated that E-glass particle can be used as partial fine aggregate replacement material as well as supplementary binding material depending on its particle size. Particles with size smaller than 75 μm could possess cementitious capability resulting from hydration or pozzolanic reaction; the coarser cylindrical might act as a potential crack arrester and inhibits the internal crack propagation [14]. Such behaviour has been confirmed by other authors [13] who demonstrated that ground waste glass containing high amount of alkali could display good performances against ASR if the fraction of fine particles can compensate the high quantity of Na_2O+K_2O.

Figure 2(b) shows that the addition of 1% superplasticizer improves compressive strength of all compositions at any ageing time. The strength of composition R1 increases from 80 MPa after 3 d of curing to 94 after 7, 114 after 28, 120 after 90, and 122 MPa after 180 d. Conversely, compositions containing 5, 10, and 20% of glass have lower strength than the reference for ageing times lower than 28 d, but their values improve faster being higher than R1 after 90 d or more. It must be also pointed out that the curve of composition V101 intercepts that of R1 shortly after 28 d and displays the best long time mechanical performance whereas those of V51 and V201 cross R after longer ageing times. Data obtained from water absorption tests (also displayed in Figure 2(b)) are in agreement with the corresponding compressive strength data, being low in materials with high strength and high in those having reduced strength levels. More in detail, it can be observed that materials with composition R1 display values between 4.7% (after 3 d of curing) and 3.1% (after 180 d) whereas these with V101 between 5.3 (after 3 d) and 2.1% (after 180 d).

Due to the irregular particle shape, blends containing glass had similar but not the same rheological behaviour as the reference compositions. Pastes workability was determined by the slump test, and pastes were defined of right workability when cake width fell into the range of 200 ± 20 mm. However, pastes containing 10 or 20 wt% of glass gave slump cakes with size close to the inferior limit, being the corresponding slurries of relatively lower fluidity with respect to the reference compositions, in agreement with the results obtained by other authors [13, 14], and the resulting hydrated materials showed a slightly higher residual porosity. The addition of superplasticizer, by improving workability, has a beneficial effect on rheological behaviour and also on the final residual porosity of compositions containing 10 or 20 wt% of waste glass.

Figure 2(c) reports the curves obtained after testing materials with 2% of superplasticizer. Compositions R2, V52, and V102 have the same strength after 28 d and similar strength trend, whereas V202 displays a lower 28 d strength, but a faster increase; the 180 d strength of all compositions are concentrated around 137 MPa. Also in this set of materials, data obtained from water absorption tests (also displayed in Figure 2(c)) are in line with the corresponding compressive strength levels. In detail, data obtained after 3 d of curing range from 5.3 for composition V202 to 3.8 for V102, whereas those acquired after 180 d are all around 2.3%; intermediate curing times give rise to materials with intermediate water absorption levels.

It must be pointed out that water absorption is not porosity but is related to the open porosity. It means that a body contains not only open porosity but also closed porosity. The water absorption test provides access to the open but not to the closed porosity which remains undetermined together with materials total porosity. However, we would like to point out the strict relationship between compressive strength and water absorption data so that their trend helps to explain material's behaviour.

XRD analysis of the hydrated samples acquired after 28, 90, and 180 d did not reveal substantial differences between reference compositions and mixed CRT glass containing materials. For comparison, the present article reports (see Figure 3) the patterns acquired on samples R1 and V201, which have a wide difference of composition and compressive strength after 90 or more days of curing. Patterns are similar, and the same phases can be identified in both materials, that is, portlandite, $Ca(Mg_{0.67}Fe_{0.33})(CO_3)_2$, calcite, and quartz. The presence of hydrated phases is not documented by this type of investigation probably due to their amorphous or cryptocrystalline nature [20, 22, 23]. However, it is possible to emphasize the presence of three small peaks which clearly appear in R and not in V201; in Figure 3 they are indicated by an arrow and may be attributed to the presence of residual calcium silicates. One could be led to infer that, in R, this phase is still present after 90 d, whereas in V201 it is almost completely consumed, thanks to the presence of the glass small particles.

The different mechanical behaviour of glass containing materials with respect to the reference glass free compositions may, moreover, be explained by the contribution to the hydration phenomena of the glass particles during mortar curing; such information could be supplied by SEM investigation, but it is necessary to start from the assumption that the coarse glass particles should contribute in a different mode with respect to the smaller ones. The SEM analysis has been made over all the samples, considering ageing time coupled with materials composition, but for brevity, only the most representative images have been reported in the present paper. Figure 4(a) shows a SEM image of the fracture surface of a sample with composition V201 after 90 d of curing. It is possible to observe that the large glass particle (dark) is strictly entrapped by the cementitious matrix; the interface appears well defined with no voids. The EDXS analysis of the particles showed that the glass mainly contains SiO_2 and BaO, but Na_2O, K_2O, and Al_2O_3 are also present; the cementitious zone mainly contains CaO accompanied by smaller amounts of SiO_2, MgO, Al_2O_3, and Fe_2O_3. It is also possible to speculate the development of a hypothetic hydrated phase containing SiO_2 (10 wt%), CaO (38%), CO_2 (40%), Al_2O_3 (3.5%), and BaO (5%) since EDXS analysis of this composition has revealed a small lump which appears well stuck to the large glass particle and is highlighted by an arrow. This hypothesis is, however, not confirmed by other investigations and, at this point, must be considered speculative.

The effect of small glass particles on the materials microstructure can be observed in Figure 4(b) where the presence of well-developed silicate hydrated crystal clusters are visible [10, 11, 18, 24]. The EDXS analysis revealed that smooth particles are CRT glass with compositions SiO_2 (62.9 wt%), CaO (1.5%), PbO (6.7%), Al_2O_3 (1.3%), Na_2O (5.4%), K_2O (6.2%), and BaO (16%) whereas the surrounding matrix and elongated crystals were detected as containing, respectively, SiO_2 (19.4 mol%), CaO (61.4%), MgO (1%), Al_2O_3 (3.2%), K_2O (8.7%), Fe_2O_3 (6.3%), SiO_2 (37.2%), CaO (58.4%), MgO (1.7%), and Al_2O_3 (2.7%). Such well-developed silicate hydrated crystal clusters were observed only around the small glass particles and not in the reference samples thus confirming that particles with size smaller than 75 μm could possess cementitious capability resulting from hydration or pozzolanic reaction concurring to limit ASR. As a consequence, provided that curing time is sufficient, the inevitable pores which are formed in the c/a matrix during mortar production may be filled with silicate hydrated crystals with a high shape ratio which interlock the surrounding material promoting the development of densely packed structures, raising compressive strength and reducing materials' permeability.

Materials aged for 180 d were also submitted to thermo gravimetric analysis (TGA) which did not supply any further information about the different behaviour between reference and glass containing compositions; consequently, TGA graphics are not specified in the present communication. It can also be pointed out that after 28 d of exposure to a moist environment, the change in length of the samples is always below the normally accepted value of 0.05% suggested in ASTM C33.

The ICP analysis made on the solutions obtained from the release tests in water of samples R and V20 is displayed in Table 3 which shows, in accordance with the work of other researchers, low elution of hazardous elements from the mortar samples containing waste materials [25]. In Table 3, only data resulted from samples containing the highest amount of mixed CRT glass and a high level of water absorption are reported since all the other compositions showed lower quantities of released hazardous elements. Some elements, such as Ca, K, Na, and S are, conversely, present in non negligible amounts, but their presence is not considered a warning parameter by most of the standard release tests. The elution release test used in the present study is not a codified test, as presently is not established leaching test for mortars or concretes containing hazardous elements, but it is indicative of the possible environmental compatibility of the materials produced. However, for safety, the Pb and Ba release from materials with composition V20 aged 28 d was also accessed by the TCLP test which showed 2.40 and 1.85 mg L^{-1}, respectively, which are far from the established limits of 5 and 100 mg L^{-1}, respectively, and in sufficiently good agreement with data reported by other authors [15–17]. It must be finally pointed out that TCLP tests are mandatory when hazardous waste materials need to be managed or disposed of to landfill, but equivalent tests on industrial products containing the same waste are presently missing. The authors of the present research therefore suggest the development of standards leaching tests to be used with monolithic materials containing waste hazardous components (i.e., ASTM or others for mortars or concretes).

TABLE 3: More abundant elements (μg Kg^{-1} = parts per billion) revealed by the ICP on the solutions obtained from the water absorption test of samples with composition R and V20. Elements not reported were determined in quantity lower than 25 ppb.

Sample name	Mg	Al	Ca	Si	Na	K	Fe	Ba	Sr	Pb	S
R	2311	3100	22005	786	16420	11003	417	<25	114	<25	19990
V20	3409	3333	16871	3956	20097	16294	390	727	326	799	21388

9.4 CONCLUSIONS

In the present research, the production of stable mortars was carried out using a commercial OPC, ground waste CRT glass, natural aggregate, and water; the addition of superplasticizer was also investigated. Mortars were produced using a fixed c/a ratio (1/3), whereas milled CRT glass was added in different proportions as well as an amount of superplasticizer.

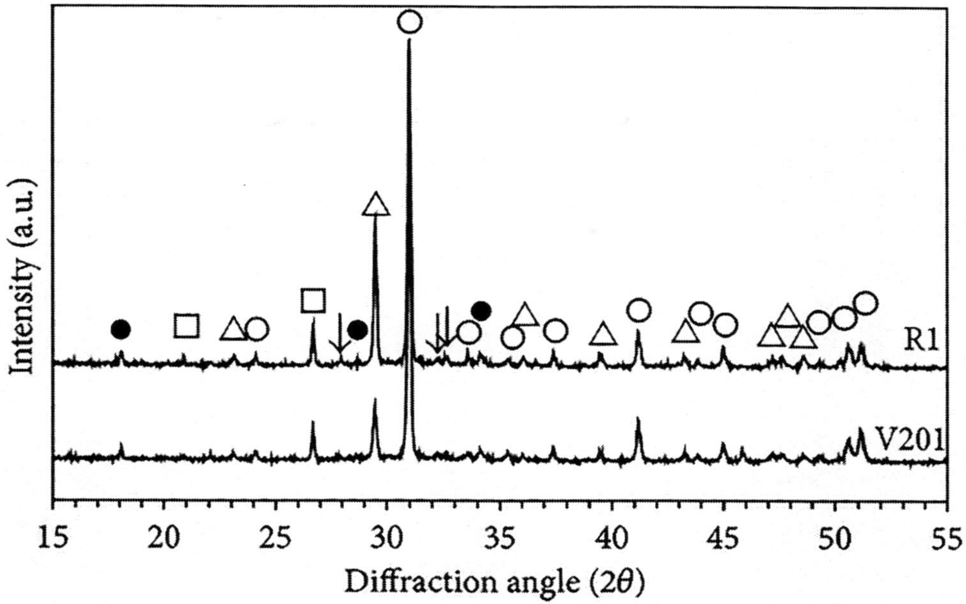

FIGURE 3: X-ray diffraction patterns between 15 and 55° of the compositions R1 and V201. The phases are identified by the following symbols: (•) portlandite; (○) $Ca(Mg_{0.67}Fe_{0.33})(CO_3)_2$; (Δ) calcite; (□) quartz; peaks due to residual calcium silicates compounds are highlighted by arrows.

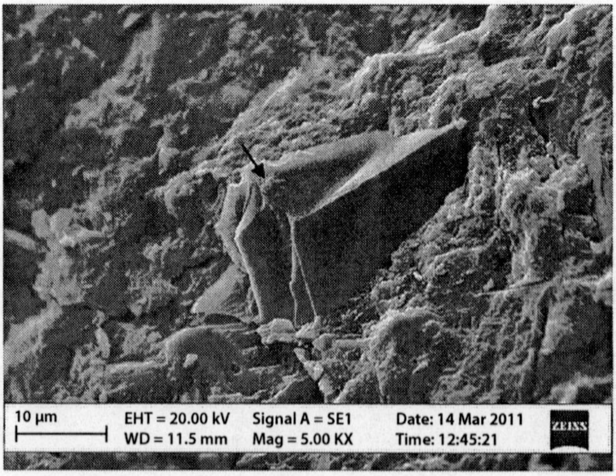

A

B

FIGURE 4: SEM micrographs showing the fracture surface of a sample with composition V201 after 90 d of curing. (a) a great glass particle (dark) is strictly entrapped by the cementitious matrix: the interface appears well defined with no voids; (b) the presence of small glass particles leads to the development of long hydrated silicates crystals.

The following important conclusions were derived from the study.

1. All hydrated materials displayed high compression strength after 3, 7, 28, 90, and 180 d of curing in a moist environment.
2. Glass containing samples showed a more rapid increase of strength with respect to the reference compositions when subjected to long-term ageing.
3. Materials with a s/c ratio of 1 showed the best overall behaviour.
4. The addition of more than 10 wt% of CRT glass powder did not lead to the production of materials with the best mechanical performances.
5. The results obtained in the present research are reasonably due to the favourable influence of the small glass particles which interact with the hydraulic phases promoting pozzolanic reaction, limiting ASR and increasing the amount of hydrated silicates produced during long term ageing.
6. The ICP analysis made on the solutions obtained from the release tests in water confirms the low elution of hazardous elements from the materials produced and therefore their possible environmental compatibility.

REFERENCES

1. C. Lee and C. Hsi, "Recycling of scrap cathode ray tubes," Environmental Science and Technology, vol. 36, no. 1, pp. 69–75, 2002.
2. S. Mostaghel and C. Samuelsson, "Metallurgical use of glass fractions from waste electric and electronic equipment (WEEE)," Waste Management, vol. 30, no. 1, pp. 140–144, 2010.
3. F. Andreola, L. Barbieri, A. Corradi, and I. Lancellotti, "Cathode ray tube glass recycling: an example of clean technology," Waste Management and Research, vol. 23, no. 4, pp. 314–321, 2005.
4. F. Andreola, L. Barbieri, E. Karamanova, I. Lancellotti, and M. Pelino, "Recycling of CRT panel glass as fluxing agent in the porcelain stoneware tile production," Ceramics International, vol. 34, no. 5, pp. 1289–1295, 2008.
5. A. M. Neville and J. J. Brooks, Concrete Technology, Longman, London, UK, 1990.
6. M. N. Haque, "Strength development and drying shrinkage of high-strength concretes," Cement and Concrete Composites, vol. 18, no. 5, pp. 333–342, 1996.

7. C. Polley, S. M. Cramer, and R. V. De La Cruz, "Potential for using waste glass in portland cement concrete," Journal of Materials in Civil Engineering, vol. 10, no. 4, pp. 210–219, 1998.
8. R. Dhir, T. Dyer, A. Tang, and Y. Cui, "Towards maximising the value and sustainable use of glass," Concrete, vol. 38, no. 1, pp. 38–40, 2004.
9. E. A. Byars, B. Morales-Hernandez, and Z. HuiYing, "Waste glass as concrete aggregate and pozzolan Laboratory and industrial projects," Concrete, vol. 38, no. 1, pp. 41–44, 2004.
10. V. Corinaldesi, G. Gnappi, G. Moriconi, and A. Montenero, "Reuse of ground waste glass as aggregate for mortars," Waste Management, vol. 25, no. 2, pp. 197–201, 2005.
11. I. B. Topçu and M. Canbaz, "Properties of concrete containing waste glass," Cement and Concrete Research, vol. 34, no. 2, pp. 267–274, 2004.
12. V. Ducman, A. Mladenovič, and J. S. Šuput, "Lightweight aggregate based on waste glass and its alkali-silica reactivity," Cement and Concrete Research, vol. 32, no. 2, pp. 223–226, 2002.
13. A. Saccani and M. C. Bignozzi, "ASR expansion behavior of recycled glass fine aggregates in concrete," Cement and Concrete Research, vol. 40, no. 4, pp. 531–536, 2010.
14. C. H. Chen, R. Huang, J. K. Wu, and C. C. Yang, "Waste E-glass particles used in cementitious mixtures," Cement and Concrete Research, vol. 36, no. 3, pp. 449–456, 2006.
15. D. Kim, M. Quinlan, and T. F. Yen, "Encapsulation of lead from hazardous CRT glass wastes using biopolymer cross-linked concrete systems," Waste Management, vol. 29, no. 1, pp. 321–328, 2009.
16. T. Ling and C. Poon, "A comparative study on the feasible use of recycled beverage and CRT funnel glass as fine aggregate in cement mortar," Journal of Cleaner Production, vol. 29-30, pp. 46–52, 2012.
17. T.-C. Ling and C.-S. Poon, "Feasible use of recycled CRT funnel glass as heavyweight fine aggregate in barite concrete," Journal of Cleaner Production, vol. 33, pp. 42–49, 2012.
18. J. Péra, S. Husson, and B. Guilhot, "Influence of finely ground limestone on cement hydration," Cement and Concrete Composites, vol. 21, no. 2, pp. 99–105, 1999.
19. H. Qasrawi, F. Shalabi, and I. Asi, "Use of low CaO unprocessed steel slag in concrete as fine aggregate," Construction and Building Materials, vol. 23, no. 2, pp. 1118–1125, 2009.
20. M. Collepardi, Scienza e Tecnologia del Calcestruzzo, Hoepli, Milan, Italy, 3rd edition, 1991.
21. M. Husem, "The effects of bond strengths between lightweight and ordinary aggregate-mortar, aggregate-cement paste on the mechanical properties of concrete," Materials Science and Engineering A, vol. 363, no. 1-2, pp. 152–158, 2003.
22. N. F. W. Taylor, Cement Chemistry, Thomas Telford, London, UK, 2nd edition, 1997.
23. P. C. Hewlett, Lea's Chemistry of Cement and Concrete, Arnorld, London, UK, 4th edition, 1998.

24. C. Shi and K. Zheng, "A review on the use of waste glasses in the production of cement and concrete," Resources Conservation and Recycling, vol. 52, pp. 234–247, 2007.
25. R. Siddique, "Use of municipal solid waste ash in concrete," Resources, Conservation and Recycling, vol. 55, no. 2, pp. 83–91, 2010.

CHAPTER 10

Challenges and Alternatives to Plastics Recycling in the Automotive Sector

LINDSAY MILLER, KATIE SOULLIERE, SUSAN SAWYER-BEAULIEU, SIMON TSENG, AND EDWIN TAM

10.1 INTRODUCTION

The use of lightweight plastics and composite materials in the automotive industry has been increasing in recent years due to legislative and consumer demands for lighter weight, fuel-efficient vehicles. The use of these materials has been credited with lowering the average vehicle weight by 200 kg [1]. In some cases, plastics are replacing heavier ferrous materials whereas, in other cases, plastics and composites are being added for consumer comfort purposes. In addition to being lightweight, these materials are also durable and easily molded. Substituting heavier materials with plastics leads to an overall weight reduction, with a 10% weight reduction resulting in a 3% to 7% improvement in fuel efficiency [2]. However, the increasing use of plastics shifts the environmental burden from the use phase of an automobile (emissions reduction) to the end-of-life vehicle (ELV) stage (materials disposal). Whether light-weighting results in an overall reduction in environmental impacts, may be soon a debatable question. A previous study concluded lighter weight vehicles show improved environmental performance even with 100% landfilling of plastic parts [3]. However, another study concluded that the environmental benefit of light-weighting would break even with negative ELV environmental impacts after approximately 132,000 km of vehicle travel [4]. Even with the use phase emissions reductions, it is critical to understand and address the ELV scenario if the use of plastics in the automotive sector can be truly considered as a sustainable option.

Plastics and composites recycling in the automotive industry is complex and challenging. Although simple plastic products (e.g., water bottles, food containers) are readily recyclable, plastics and composites in automotive applications are heterogeneous, have strong connections to other plastics, and are thus difficult to liberate for recycling [5]. Thermoset materials present a further challenge since they cannot be melted down and recycled due to their permanent cross-link structure. Even when a material can be recycled, it is often still landfilled because it cannot actually be physically recovered [5]. The placement of foam, for example, is typically in an area of a vehicle where it cannot be readily accessed and then separated from other materials: it will eventually be contaminated with other materials (e.g., fluids). Furthermore, there are additional obstacles blocking the recycling routes such as a lack of technology and market for recyclates. Lastly, next generation materials, such as carbon fiber, further complicate recycling due to their inherent complexity [6]. It is generally found along with other materials, and is difficult to separate. Bio-based plastics present an interesting alternative to petroleum plastics but the technology needs maturity before implementation.

For a spent automobile, the current ELV recovery and recycling process consists of dismantling, de-pollution and shredding, physical and mechanical treatment of shredder residue (SR), and treatment of SR by energy recovery [7]. Presently, plastics and composites contribute to SR, and as the use of these materials increases, so does the amount of SR generated. The percent plastics by mass in an average vehicle has gone from 6% in 1970 up to 16% in 2010 and is expected to reach 18% in 2020 (see Figure 1) [8]. The automotive industry accounts for a significant percentage of plastics demand with estimates ranging up to 30% and on the rise [9,10].

Identifying possible improvements and alternatives in the plastics recycling chain is of utmost importance in Europe given upcoming legislation that will require 95% of a vehicle to be processed for "reuse and recovery" and 85% be processed for "reuse and recycling" by 1st January 2015 [11]. This legislation limits the amount of the ELV that can be dealt with through recovery and encourages recycling whenever environmentally viable. It is expected that by increasing the recycling of plastics from SR, an additional 6%–10% of the entire ELV mass can be recycled, therefore making this a key initiative towards meeting legislative requirements [12]. While North America does not have similar legislation, the global nature of the automotive industry has resulted in many manufacturers adopting global initiatives towards regional environmental requirements.

In summary, the recycling of plastics and composites from complex, durable products is limited by technological and economical restraints. The challenges in recycling plastics and composites from ELVs stem from a lack of market for recyclates, a lack of infrastructure, economics, knowledge gap, and heterogeneous mixes of plastics. This paper examines the challenges of plastics recycling in the North American automotive industry and suggests the following alternatives to overcome these challenges: (1) pursue plastics reuse and refurbishment; (2) pursue recycling; (3) pursue recovery without segregation (energy recovery); (4) move to renewable plastics; and (5) move away from plastics.

10.2 CHALLENGES AND OBSTACLES ASSOCIATED WITH PLASTICS RECYCLING

In order to meet the requirements for CO_2 emission reductions, vehicles are being light weighted with more plastics substituting for heavier materials, such as steel, where possible. CO_2 emissions originate predominantly from the use phase, with only ~1% being attributed to the recycling and waste stage (see Figure 2) [8]. In this regard, plastics are being credited for lessening environmental impacts. However, only a few plastic parts actually are recycled; typically these are the fascia (or "bumpers"), dashboards, and battery casings [1].

FIGURE 1: Change in vehicle composition from 1970 to 2010 [8].

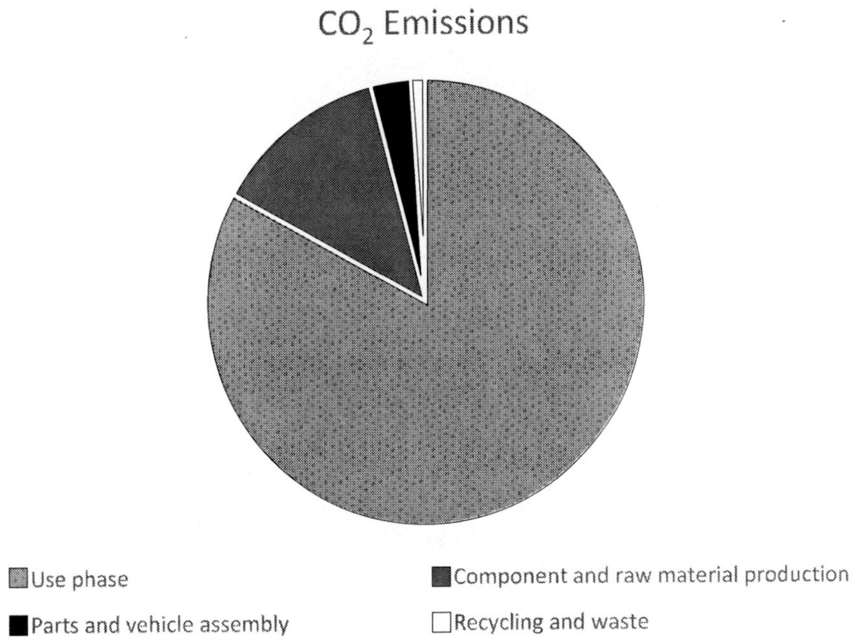

FIGURE 2: CO_2 emissions during various life stages of an average vehicle (1000 kg) [8].

An important distinction to make in the discussion of difficulties surrounding plastics recycling is that between thermosets and thermoplastics. Thermoset materials cure into a given shape through the application of heat. Curing results in permanent cross-links resulting in a high degree of rigidity, however, changes the material permanently. Thermoset materials will not remelt or regain processability. The only option for mechanical recycling is to pulverize these materials for reuse as fillers [13]. In contrast, thermoplastic materials become pliable when heated, allowing them to be moulded, but they do not set. These materials typically begin in pellet form and are heated and moulded. As the material cools it will harden, but no curing takes place and cross-links are not formed. This allows for thermoplastic materials to be reprocessed many times although continual recycling will result in degradation. Although there are more options for recycling of thermoplastics, thermosets can be used for other applications such as in the case of polyurethane foam (PU) foam which is commonly shredded and used as carpet underlay.

There are several obstacles to recycling the types of materials that are commonly used in automotive applications. Plastics that are commonly used in automotive applications include: polyurethane foam (PU) polypropylene (PP), polyethylene terephthalate (PET), polyethylene (PE), polyvinyl chloride (PVC), and acrylonitrile butadiene styrene (ABS). Polyurethane foam, for example, has been gaining widespread use in the automotive industry due to its lightweight, mouldable, and durable properties. It is also a useful noise dampener. These desirable properties are responsible for the increase in usage of this material within vehicles despite it being a thermoset material and other materials being more easily recycled [14]. Polyurethane is commonly found in automotive seating applications as well as within the interior and under the hood. Unlike some other thermoset materials, polyurethane has the advantage of being relatively easy to convert back into its original monomer [15]. Therefore, this, in theory, is recyclable and several technologies exist for this purpose. In reality, the location of the foam within the vehicle is not easily accessible and is often contaminated. Furthermore, it is currently not economically viable for dismantlers to segregate this inexpensive material from the end of life vehicles. Some other vehicle components that are commonly made of plastics include drink trays (PP), armrest finishers (PP/ABS), and seatbelts (PET). Similar to PU foam components, these often do not make it to recycling due to inaccessibility and lack of economic incentive at the dismantling stage.

Carbon fibre reinforced plastic is another composite that is theoretically recyclable yet is often landfilled. This material is also used in the automotive industry although usually reserved for high end vehicles and racing cars due to its higher costs. Recycling carbon fibre reinforced plastic is inherently difficult due to its complex composition and cross-linked nature [16]. Additionally, it is usually found in combination with other materials such as metal fixings and hybrid composites, and as a result, it is difficult to separate them for subsequent recycling.

PLA is a renewable bio-plastic made from the fermentation of starches utilized in automotive interior components such as dashboard trim and spare tire cover. Current methods for PLA recycling are mechanical and chemical reprocessing, and unlike conventional plastics, composting. Mechanical processing has shown to reduce the physical properties at each application [17]. For mechanical processing, a homogeneous (sorted) feedstock is necessary to preserve the physical properties. For composting, PLA is a biodegradable plastic under industrial composting conditions according to EN 13423, ASTM D6400, and D6868. The challenges associated with PLA recycling are identifying and then sorting them. With respect to infrastructure, the challenges are cost and implementation. Similar to other commodity plastics, two-thirds of the total financial cost in plastic recycling is incurred by collection and sorting [18].

Several common obstacles can be established based on the discussion to date.

10.2.1 A LACK OF MARKET FOR RECYCLATES

The benefit of segregating and targeting a material for recycling is diminished if there is no market for the material once it is recycled. The cost of some of these materials is already low; for example, polyurethane foam. Furthermore, even for meltable thermoplastics the mechanical properties of these materials may be altered during the recycling process, rendering them less suitable to be recycled for their original purpose. Materials used within the vehicle are also often times contaminated, such as PU foam used as an engine compartment would likely become contaminated with oils. Similarly, carbon fibre undergoes physical changes as a result of the recycling process preventing its reintroduction as a direct substitute for virgin fibres. For these reasons, these materials are more likely to be down-cycled than truly recycled, especially in the case of thermoset materials. PU foam for example, has been down-cycled for use in carpet underlay whereas carbon fibres have been in the construction industry as fillers for artificial woods and asphalt [19]. The use of foams and composites is increasing; however, the market size for these rebounded applications is not large enough to accompany the recycled materials.

10.2.2 A LACK OF INFRASTRUCTURE

Plastic manufacturers recycle their own scrap materials in house. Once the plastics leave the manufacturers for use in their end application, such as automotive components, the recycling process is complicated by a lack of infrastructure. Current automobile recycling infrastructure consists of a dismantler, shredder, and non-ferrous operator. The dismantler, after the de-pollution step, removes components with sufficient market value such as aluminum rims and catalytic converters, for reuse, remanufacturing or recycling. Parts recovery values in the industry will vary depending on the condition, makes, models, and ages of the ELVs processed, as well as the market demand for the particular part and assembly types. In general, this can amount to as little as 5% to as much as 42% of the vehicle mass [7,20]. The remainder is sent to a shredder facility. Post-shredding, mechanical and magnetic separation processes allow for the recovery of ferrous metals.

Approximately 15%–25% of the mass is leftover and ends up at a landfill in the form of shredder residue (SR). Foams, plastics, and polymer composites typically end up as SR. Upcoming European legislation will force a reduction to 5%, which will require further segregation and recycling of the plastic materials. In order to target these materials, a change in infrastructure will be required. In addition to meeting more stringent regulations is the concern over the changes in SR composition. It is expected that new vehicle composition will reach 10%–15% plastic materials compared to the current 6%–8% [21].

10.2.3 ECONOMICS

The previous two obstacles are both largely based on economics. Without a sound market for recyclate, it is not economical to recycle these materials. Dismantling operations are based on recovering materials for profit, and cannot afford to separate low values materials. Furthermore, it is not economical to improve or develop infrastructure for new technologies if no future revenue is expected to be generated. In the absence of a real market for the reuse or recycle of these materials, the economic situation will likely only be addressed through legislative efforts. One exception are carbon fibre materials, which have high manufacturing and material costs as well as disposal costs and could benefit economically from recycling practices.

10.2.4 MINDSET AND KNOWLEDGE GAP

The mindset of the public is that plastic materials are easily recycled based on the public knowledge of recyclable products such as water bottles, which are typically made from a single thermoplastic material type, not bonded to any other items, and are easily collected and transported. Manufacturers and consumers also favour plastics to reduce the overall vehicle weight and thereby lower fuel consumption and subsequent emissions. Although plastics are lighter weight than some ferrous materials, replacing parts with plastics will not always yield a positive environmental impact over the life of a vehicle. This poses a knowledge gap as manufacturers may select a plastic component partially based on its ability to be recycled. This was demonstrated using a case study of material selection for an engine cover [22]. Although PU foam was selected for its recyclability, the end-of-life circumstances were such that no matter what material was selected for this part, it would be landfilled. In this case, the best material selection would have been the lightest weight component to reduce the use phase environmental impacts. However, as plastic usage continues to rise, there will likely be consequences at the end-of-life stage, such as a percentage increase in landfilling, if sustainable solutions cannot be achieved.

10.2.5 HETEROGENEOUS MATERIALS AND HETEROGENEOUS MIXTURE WITHIN AN APPLICATION

The heterogeneous nature of these materials makes it difficult to recover without losing some of their original mechanical properties. Moreover, in addition to the heterogeneity of these materials, they are often coated. Coating materials can compromise the properties of recycled plastics [23]. Perhaps even more difficult is the heterogeneous mixture of materials within an application. For example, within the seating section of a vehicle, there can be five different types of plastics and composites, some being thermoplastics and others thermosets. This complicates the recovery process since in order for it to be effective, all of these materials would have to be recovered and separated at the end of life processing. This increases the time required to dismantle a vehicle and could reduce the cost effectiveness of the process if commercial materials recyclers having suitable materials separation technologies are not readily available. Another option is to separate materials post shredding; however, unlike steel and light metals, plastics are not easily separated from one another due to their overlapping physical properties [24].

As a result of these obstacles, what is technically possible in terms of recycling are often times not practically feasible. The reality is that most often the plastics and composites used within the automotive sector end up as automotive shredder residue and have consequently been landfilled. With landfilling no longer an option in Europe, there is a pressing need to develop feasible solutions.

TABLE 1: Examples of parts assemblies having significant non-metallic materials content [20].

Part Assembly	Average		Range	
	Metals (% Weight of part)	Non-Metals (% Weight of part)	Metals (% Weight of part)	Non-Metals (% Weight of part)
Front door assembly	70%	30%	66%–72%	28%–34%
Rear door assembly	69%	31%	66%–74%	26%–34%
Steering wheel	60%	40%	55%–65%	35%–45%
Steering column	80%	20%	77%–83%	17%–23%
Dash assembly	44%	56%	28%–54%	46%–72%
Seat assembly	66%	34%	21%–81%	19%–79%

10.3 ALTERNATIVE SOLUTIONS FOR PLASTICS RECYCLING

10.3.1 ALTERNATIVE 1—PURSUE PLASTICS REUSE/ REFURBISHMENT

According to Sawyer-Beaulieu's 2009 study, as much as 12% on average by weight of end-of-life vehicles (ELVs) entering the dismantling processes are recovered and directed for either, reuse, remanufacturing or recycling, including the recovered fluids. Of this 12%, more than 80% of these recovered end-of-life materials and components (almost 10% by weight of the processed ELVs) are parts and parts assemblies directed for reuse, remanufacturing and recycling. Parts and assemblies directed for reuse and remanufacturing were almost 6% weight of the ELVs processed and represented hundreds of different part and assembly types [20].

The materials compositions of these reusable and remanufacturable parts and assemblies vary in complexity. These include: (1) parts composed principally of metals (e.g., engines, transmissions, AC compressors, steering gears, radiator supports, alternators, starters, etc.) and very amenable to reuse and/or remanufacturing (or refurbishment); and (2) parts assemblies of relatively complex materials compositions, having significant non-metallic materials content (refer to examples in Table 1, below). Should these latter types of parts be resold for reuse, then these represent instances of plastics reuse even if it is incidental reuse of the onboard plastics. Based on literature available in the public domain, the environmental benefits of "incidental" plastics reuse, through the reuse of automotive parts or assemblies having complex materials compositions, have not been explored.

Fascia, or commonly the front and rear "bumper covers", are one of the few large, predominantly plastic automotive parts that are readily recovered by dismantlers and sold for direct reuse and refurbishment and reuse [20]. There are companies in Canada and the U.S. that refurbish and sell recovered bumper covers as alternatives to using new original equipment manufacturer (OEM) or aftermarket bumper covers. LKQ Corporation/Keystone Automotive, for example, own approximately 37 facilities dedicated to bumper cover repair in the United States, Canada and Mexico that employ approximately 600 workers [25]. Fascia refurbishment is one of the limited examples of direct reuse—or refurbishment and reuse—of automotive parts that are made principally or exclusively of plastic, likely because they are readily identifiable, isolated, and relatively easy to handle. Nevertheless, the recovery and subsequent resale of other similar plastic assemblies from automobiles or other consumer products merits serious consideration as part of the larger effort to achieve materials sustainability.

10.3.2 ALTERNATIVE 2—PURSUE RECYCLING

Technologically, recycling is possible for the plastic and composite materials used in the automotive industry. Excellent reviews of the technical aspects and options for recycling plastics and composites are available [1,15,16,26]. While recycling is practically limited by the above mentioned obstacles, it may still be a preferred option for these materials and can help to alleviate the amount of materials being landfilled if some of the obstacles can be overcome. The feasibility of recycling can be improved through a few major avenues.

10.3.2.1 IMPROVE THE QUALITY OF THE RECYCLATE

If recycling these materials is to succeed, plastic and composite waste need to be rendered into a valuable resource. This will help drive higher value applications for recyclate. In order to move in this direction, the by-products resulting from mechanical recycling processes would need to have similar properties to commercial grade plastics with respect to their type and monomer origin [26].

New innovative technologies will be paramount to improving recyclate quality. One such technology is Polyfloat®, developed by SiCon, which enables high-precision density separation of plastics from shredder residue. The process relies on a lamella separation system and is applicable to the plastics that are commonly found within vehicles [27]. Another option would be to find more uses for SR, in the event dismantling more material before shredding proves to be uneconomical [7].

The value of the recyclate derived from thermoset materials can be improved by finding applications in which the properties of this recyclate can by uniquely applied. For example, using the recyclate as a permeable core that allows it to act as a flow layer [28] as well as using it to provide damping in a composite for the purpose of noise insulation [29]. Further improvements in thermoset recycling lie in the development of new materials that have the desirable properties of a thermoset material however are recyclable at the ELV stage. Advancements have been made recently towards this improvement [30].

10.3.2.2 ESTABLISH INDUSTRY PARTNERSHIPS

The most likely route to achieving recycling success is through developing mutually beneficial partnerships where recycling benefits all stakeholders. Because recycling these materials often does not result in the direct substitution of virgin material for the part's original purpose, the next preferred solution is to find a suitable purpose for the recycled material. By establishing collaborations between waste generators and recyclers, the generators can focus on material sorting and reduction of contamination to reduce the processing burden on the recycler. An example of this is the collaboration between Boeing and BMW. With Boeing's 787 Dreamliner made of 50% carbon fibre material, recycling it at end of life is essential. BMW is working to bring two vehicles with a carbon passenger cell onto the market and therefore has a potential use for the recycled carbon fibres. In a joint venture between the two companies, infrastructure has been developed to turn the carbon fibre material into fabric, which is then processed to make body components for BMW [31]. This partnership, while nascent, serves as an example of industrial ecology to encourage other companies to seek out mutually beneficial collaborations. Since carbon fibre is also used in several sectors such as automotive, aircraft, and watercraft, recycling partnerships could be easily established if infrastructure were available.

10.3.2.3 INCENTIVES AND LEGISLATION

Overcoming the obstacles surrounding the economics of recycling will likely require government involvement and initiatives. The EU ELV legislation, for example, requires 95% of a vehicle to be processed for "reuse and recovery" and 85% be processed for "reuse and recycling" by 1 January 2015 [11]. Vehicle manufacturers are economically responsible for the ultimate recyclability of vehicles, including the end-of-life treatments, recycling, and recovery operations [32]. ELV recovery and recycling rates reported by EU member states for 2011 (refer to Figure 3) varied from about 74% to 93% for reuse and recycling and 79% to 97% for reuse and recovery [33]. It should be noted that the 2011 figures are the most current figures available. Germany (i.e., country code "DE") reported reuse and recovery rates in excess of 100% which may imply that more materials were reused and recovered than generated. This may be possible if materials were imported for reuse and recovery. Based on the 2011 statistics, 63% of EU member states have yet to attain the 85% EU reuse and recycling target and 87% to attain the 95% EU reuse and recovery target.

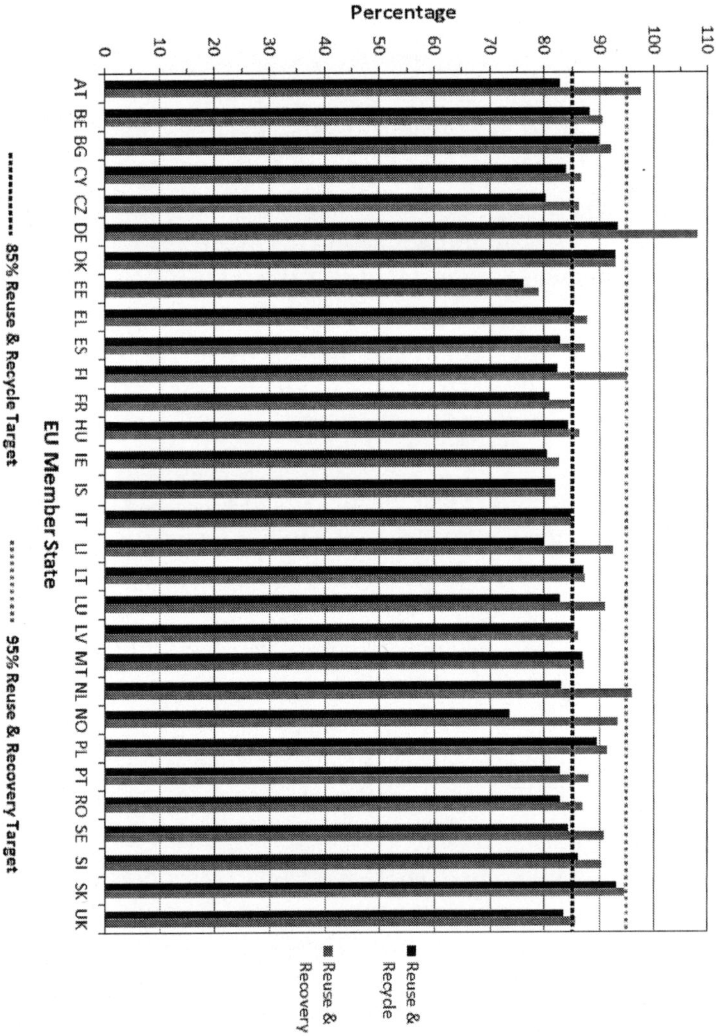

FIGURE 3: End-of-life vehicle (ELV) recovery and recycling rates reported by EU member states for 2011 [33].

In China ELVs are managed under the End-of-Life Vehicle Recycling Regulations (enacted in 2001), the Automotive Products Recycling Technology Policy (implemented in 2006) and the Regulation of Remanufacturing of Pilot Automotive Parts (enforced in 2008) [34]. Collectively they establish an ELV collection system that controls how vehicles may be used and managed at end-of-life. The system promotes improved safety by preventing the illegal refurbishment and use of old age vehicles, as well as controls how ELV-derived parts and materials may be reused. Similarly to EU's ELV Directive, China's ELV regulatory system, also sets ELV recycling targets, about 85% by 2010 (minimum 80% material recycling), about 90% by 2012 (minimum 80% material recycling) and about 95% by 2017 (minimum 85% material recycling) [34].

Japan's Law for the Recycling of End-of-Life Vehicles (enforced in 2005) was established, in part, to address the problem of illegal dumping of automotive shredder residue (ASR) and diminishing space in disposal sites. The ELV Recycling Law sets recycling targets for airbags (85%) and ASR: 30% from 2005 to 2009, 50% from 2010 to 2014 and 70% from 2015 on. Automobile manufacturers and importers are responsible for the recycling of the ELVs. Unlike the EU ELV directive, Japan's ELV Recycling Law does not stipulate recycling or recovery targets with respect to the total vehicle weight [34].

In Canada and the United States, regulation of ELV management facilities primarily focuses on business and operating practices as opposed to the regulation of the retired vehicles themselves [35]. The regulatory mechanisms (legal statutes, regulations, bylaws and voluntary mechanisms, such as best management practices (BMPs)) are applied to ELV management activities, such as, pollution prevention and control of air contaminant emissions, waste water discharges, and waster generation and disposal, as well as site use and materials storage practices [35]. The recycling of ELVs and ELV-derived parts and materials is principally a market drive system with used parts and scrap metal prices driving high recycling rates [36], which have been estimated to be as much as 86.3% by weight of ELVs processed in the [32]. Although there is no extended producer responsibility (EPR) or product stewardship legislation for ELVs managed in Canada and the United States, EPR-based initiatives have been launched for a variety of ELV-derived materials, or are under consideration (refer to Table 2).

TABLE 2. Examples of extended producer responsibility (EPR)-based initiatives for ELV-derived materials in Canada and the United States.

EPR-Based Initiative	Program	Jurisdiction	Source
Automotive mercury-containing switches	National Vehicle Mercury Switch Recovery Program (NVMSRP)	United States	[37]
	Switch Out Program	Canada	[38]
Used tires	Ontario Tire Stewardship (under Ontario Waste Diversion Act 2002)	Ontario	[39]
	Balanced Budget Act—Introduction of an environmental fee at time of new tire purchase	Quebec	[40]
Used oil and oil filters	Ontario Regulation 85/03, Used Oil Material (under Ontario Waste Diversion Act 2002)	Ontario	[41]
	Regulation respecting the recovery and reclamation of used oils, oil or fluid containers and used filters	Quebec	[40]
ELVs and ELV-derived materials	Proposed diversion of ELVs and ELV-derived materials from landfill (proposed changes under Ontario Waste Diversion Act 2002)	Ontario	[42]
	National ELV Environmental Management System (proposed)	Canada	[43]

Based on current markets and economics, any dismantling operation over and above what currently takes places will likely require a subsidy to compensate for the economic loss that would be incurred by the dismantler or recycler [43]. The introduction of incentives to push recycling may be one possible solution to promote segregation at the dismantling stage. Subsidies at the dismantling level may encourage separation of parts that are often not of significant value. For example, in the Netherlands, the dismantling, logistics, and sales chain for polyurethane foam is subsidised by Auto Recycling Netherland [43]. Subsidies could also be provided to drive research and development of the some of the technologies that are still not commercially available. Such subsidies may assist with "jump starting" the recycling process. Furthermore, introducing legislation which requires minimum recycling rates would drive the research and development of recycling technology and the development of partnerships. Legislation, such as the EU directive that 95% of vehicles disposed after 2015 must be recyclable will also help to drive research and development of technology and partnerships to achieve these milestones. Another possible avenue is to raise the cost (or "taxing") production processes which rely completely on virgin materials while providing incentives to incorporate recycled materials.

Legislation could also help with creating a market for recyclates. For example, legislation requiring manufacturers to design new vehicles with a minimum percentage of materials obtained from SR would help to develop a real market for end of life vehicle materials. Investment in infrastructure should focus on one of two avenues: targeting more complete dismantling or developing SR processing technologies such as Polyfloat® to achieve separation post-shredding. Interestingly, while the U.S. and Canada have largely resisted EU style regulations on recycling in the automotive industry, the EU regulations have spurred global, often voluntary actions related to end-of-life issues among a number of the automotive OEMs due to the interconnected and global nature of the automotive sector.

10.3.2.4 KNOWLEDGE TRANSFER (MOVE TO UNIFORM MIX OF MATERIALS)

A previous case study has demonstrated that a knowledge gap exists between what manufacturers believe are positive environmental choices and end of life reality [22]. Additionally, there is the obstacle of conveying information, such as the location of recyclable materials, from the manufacturer to the ELV dismantler. These gaps need to be addressed through collaborative efforts. One tool that has been developed is the International Dismantling Information System (IDIS) database [44]. The IDIS database provides information to assist with vehicle dismantling; specifically, it shows where materials can be found within the vehicle. This database has the potential to increase recyclability by assisting dismantlers with identifying where recyclable plastics are within the vehicle; however, there are still several limitations such as reporting inconsistencies between manufacturers.

Manufacturers also need to be aware of the limitations surrounding recycling of thermoset materials and specify these only in situations where their properties are required. Awareness of the end of life routes for each material they consider as well as of the contradiction between heterogeneity and recyclability is also important. Rather than consider whether a material is recyclable, it is more important to consider the likely end of life fate. Design for Resource Efficiency (DfRE) is a concept in which the end of life scenario is included within the design phase with regards to how the resources can be extracted from the dismantling and recycling processes [45]. This gap can be bridged through publications and conferences and through forging partnerships between manufacturers and researchers. One important factor to convey to manufacturers might be how reducing the diversity of materials during production may increase recyclability.

10.3.3 ALTERNATIVE 3—PURSUE RECOVERY WITHOUT PLASTICS SEGREGATION (ENERGY RECOVERY)

Recovering unsegregated ELV materials (including plastics) is commonplace in the metals shredding industry. Since shredding was first introduced in the early 1960s [46], the basic materials recovery technologies have matured and are in wide use. End-of-life vehicle (ELV) hulks, end-of-life large appliances (ELLAs), and construction, renovation and demolition (CRD) waste are typical shredder feed stocks. These are comminuted through large metal shredders (hammermill, typical) producing a heterogeneous mixture of metals (ferrous and non-ferrous) and non-metals (e.g., plastics, glass, textiles, rubber/elastomers, paper, wood, ceramics, etc.) [20]. From this shredded mixture, magnetic ferrous metals (cast iron, carbon steel, low grade stainless steels) are recovered using magnetic separation systems (e.g., magnetic drum, magnetic head pulley, magnetic belt separators, etc.) [47]. Non-magnetic, non-ferrous metals (aluminum, copper, zinc, nickel, high-grade stainless steels, lead, etc.) are recovered and concentrated principally using eddy current rotor separators, commonly in combination with screening devices such as trommel or vibrating deck screens to remove fines [20,48,49]. The low density, non-metallic materials are removed from the heavier, metal-rich materials using air suction and conveyed to air elutriation devices for recovery (e.g., vertical air classifiers, such as Z-box separators and air cyclone separators) [20].

The "left overs" of this process is a heterogeneous mixture of non-metallic materials with a small proportion of non-recoverable metals, and is referred to as shredder residue (SR) or shredder fluff. It is also commonly referred to as auto shredder residue, which can be misleading given that it is typically generated as a result of shredding ELVs along with ELLAs and CRD waste. SR is disposed of in landfills [20], but can be costly, particularly if it is deemed hazardous due to the presence of sufficient quantities of leachable contaminants, such as mercury (from mercury switches), polychlorinated biphenyl or PCB (from PCB components in ELLAs or CRD waste commingled with ELVs), or lead (from soldered wire connections) [20]. Alternative SR management mechanisms have been explored and include reuse, recycling and energy recovery options, such as:

- reuse of SR as landfill day cover [50,51,52];
- reuse of SR as a hydroponic garden growing medium [53];
- use of the organic portion of SR as an alternative fuel source or reducing agent in blast furnaces [50,54,55];
- use of SR as an alternative fuel and mineral feedstock for cement production [55,56];
- recycling of SR in the manufacture of composite plastic building products, e.g., plastic lumber [57,58];
- pyrolysis of SR to produce a synthetic coal/fuel product [51,55,59,60,61,62]; and
- recycling of SR plastics, involving the conversion of the plastics into low molecular weight hydrocarbons (such as via low-temperature, catalytic conversion) for reuse as chemicals or fuels [55,62,63].

The aforementioned reuse and recycling mechanisms for non-segregated SR materials have been generally limited to proposed, experimental or conditional applications. Although the above alternatives may be viable and seen as environmentally beneficial ways of reusing or recycling SR, are they sustainable?

Using SR as landfill day cover is practiced in the U.S. and Canada, but landfill facilities require prior approval to use SR as alternative day cover (ADC). Even as ADC, SR takes up landfill space and concerns about SR quality may arise, due to potential contaminants, for example, automotive fluids, PCBs, or leachable heavy metals, unless SR quality is regularly monitored [51,52].

Energy and resources are necessary both to shred and recover the materials. Pretreating the SR materials may be required (e.g., screening, air elutriation, froth flotation, etc.) to upgrade and concentrate the organic portion of SR (principally plastics) prior to it being used in energy recovery applications or recycled into manufactured plastic products [64,65,66]. The use of SR organic materials for energy recovery results in the consumption of non-renewable resources in secondary processes. If the amount of energy used to generate the SR and transform it into a useable fuel is greater than the calorific value of the SR-derived fuel, justifying SR as a fuel source may be difficult.

10.3.4 ALTERNATIVE 4—MOVE TO RENEWABLE PLASTICS

Social and economic progress of the 21st century has propelled renewable plastics to the forefront of the plastics industry. Bio-plastics are now considered a viable alternative to conventional petroleum based plastics. Bio-plastics, even as an emerging industry, has many virtues worthy of the transition away from petroleum based plastics in the current literature [67,68]: the two most influential factors are renewability and biodegradability.

Renewability with relation to bio-plastics can be defined as plastics manufactured from sources that are replenished naturally on an anthropological time frame. Some examples of renewable sources of bio-plastics are corn, sugar cane and algae. In the short and medium term, the price fluctuation due to the finite supply of fossil oil can be minimized by utilizing renewable biomass as raw materials [69]. In the long term, the development of bio-plastic is driven by the acknowledged, definite supply of fossil oil [70]. By making the transition towards utilizing bio-plastics, social concerns, production cost and environmental impacts can be lowered.

Biodegradability is commonly used to define a product's ability to completely disintegrate chemically and biologically; however, there are many kinds of biodegradability not well known among the general public. For bio-plastics, degradability under commercial composting conditions and household composting conditions are expected. EN13432 and ASTM

D6400 are the two standards that define the industrial composting [71], while household composting conditions are not standardized. Similar to renewability, the biodegradability of bio-plastics reduces the environmental impacts of plastic products and moves the plastic industry overall to a greater level of long-term sustainability.

For automotive applications, bio-plastics are currently limited to interior trims and non-structural components with ongoing research to improve the material properties for universal application. The mature production infrastructures of petroleum plastics, however, remain obstacles to bio-plastics becoming main stream commodity plastics. Future social pressures and subsequent market trends and even government legislation will dictate the eventual route of bio-plastics progression.

10.3.5 ALTERNATIVE 5—MOVE AWAY FROM PLASTICS

When plastics were first introduced into the automotive industry, their light weight and mouldability were seen as improvements over metals for certain applications. However, while plastics may reduce the overall weight of a vehicle, they are not readily recoverable and recyclable compared with metals [34]. The initially perceived benefit to switching to plastics is being questioned. A comprehensive life cycle analysis might expose metals as having a smaller environmental impact associated with their tertiary life cycles compared to the end-of-life scenario for currently unrecoverable plastics [72].

Whether it would be beneficial to reduce plastics use and re-emphasize metals use in automotive applications depends on a variety of factors. Ferrous metals are readily recyclable from ELVs using magnetic separators, while plastics add to the complexity within the shredding residue [34]. While dismantling ELVs can further reduce the environmental impacts, not all plastics can be dismantled from the vehicle [72]. Ironically, using metal clips and screws instead of typical glues to join plastics together may improve separation during recovery [73]. Nevertheless, metals are generally heavier than plastics, and results in greater impacts during the use-phase from an LCA perspective. Further investigations from an LCA perspective contrasting the use-phase and end-of-life phase from metals compared with plastics could identify whether moving away from plastics would be a worthwhile alternative [74].

10.4 RECOMMENDATIONS AND CONCLUSIONS

Five alternatives were discussed in relation to the options for plastics and composites from ELVs. One of the first, critical questions is whether or not conventional plastics are the best choice for the application being considered. In some cases, bio-plastics or metals may provide superior functionality and environmental performance over the life cycle of the vehicle given the challenges of recovering plastics. The mindset that plastics are a preferred material choice needs to be challenged.

In cases where conventional plastics are used, the best recovery alternative is likely a combination of the reviewed options; for example, combining recycling and energy recovery steps. Segregating the larger, cleaner material pieces for recycling and sending the remainder for energy recovery is a common choice; however, future efforts should look to increase the proportion destined for recycling. For thermoset materials, emerging techniques are showing promising results towards the development of new materials that can have the desirable properties of thermoset plastics while being recyclable at the ELV stage. Several of the alternatives could exist in combination to comprehensively address the challenges of recycling with the overall goal of improving the economics and efficiency of identifying, separating, and recovering plastics.

In the absence of major investment in infrastructure and new technology, an alternative may be to promote increased uses of SR. Post-shredding SR treatment technologies can play an important role in increasing overall recycling rates. Although these technologies exist, they are currently available on a limited commercial scale [66]. Technologies are available; however, it will take incentives and legislation to commercialize them. Legislation could advance this initiative by requiring manufacturers to incorporate SR materials into new design.

Subsidies and legislation may play a critical role in addressing the abovementioned obstacles. A major limitation of sustainable recycling of these products is cost. Incentives could propel some of these alternatives forward. Providing penalties for waste generators and credits for recyclers and material recycling initiatives is one option, while subsidies could be provided at various levels.

Incorporating recovery principles into the product design stage is critical. Firstly, informing manufacturers of the likely EOL fate of their materials is an important component. Also, the need for manufacturers to adapt the mindset of sustainable design is imperative. This can be achieved by reducing the number of types of materials that are used in order to streamline segregation activities, or alternatively, selecting materials with easy-to-recover properties in applications where plastics recovery will be difficult.

Establishing applications and sound markets for recyclates through collaborative efforts and partnerships will be key to closing the materials reuse loop. Where energy recovery is the only option, waste to energy practices can be implemented.

In conclusion, plastics play an important role in the automotive industry, but there are obstacles to overcome in order to ensure that they are sustainable. Pursuing recycling in the automotive industry will require improvements in the quality of the recyclate, establishing industry partnerships, incentives and legislation, and knowledge transfer to industry stakeholders. Implementing some of the innovative alternative SR management techniques will require significant, further research to determine if they are sustainable practices. Moving to renewable plastics will be determined by industry maturity and progression. Switching away from plastics and back to metals will require proper analysis between the impacts from the use-phase and end-of-life phase from an LCA perspective. Several options exist for the EOL fate of plastics but none truly stand out as the preferred alternative: future research is critical to determining what the ideal combination of alternatives might be for long term sustainability. Finally, many of the issues raised and lessons learned to date from the use of plastics in vehicles may be applicable to the use of plastics in other increasingly complex, durable consumer and industrial items.

REFERENCES

1. Yang, Y.; Boom, R.; Irion, B.; van Heerden, D.-J.; Kuiper, P.; de Wit, H. Recycling of composite materials. Chem. Eng. Process. 2012, 51, 53–68.
2. McAuley, J.W. Global sustainability and key needs in future automotive design. Environ. Sci. Technol. 2003, 37, 5414–5416.
3. Leduc, G.; Mongelli, I.; Uihlein, A.; Nemry, F. How can our cars become less polluting? An assessment of the environmental impact potential of cars. Transp. Pol. 2010, 17, 409–419.
4. Duflou, J.R.; de Moor, J.; Verpoest, I.; Dewulf, W. environmental impact analysis of composite use in car manufacturing. CIRP Ann. Manuf. Technol. 2009, 58, 9–12.
5. Buekens, A.; Zhou, X. Recycling plastics from automotive shredder residues: A review. J. Mater. Cycles Waste Manag. 2014, 16, 398–414.
6. Van Acker, K.; Verpoest, I.; de Moor, J.; Duflou, J.-R.; Dewulf, W. Lightweight materials for the automotive: Environmental impact analysis of the use of composites. Revue Metal. 2009, 106, 541–546.
7. Vermeulen, I.; van Caneghem, J.; Block, C.; Baeyens, J.; Vandecasteele, C. Automotive Shredder Residue (ASR): Reviewing its production from end of life vehicles (ELVs) and its recycling, energy, or chemicals' valorization. J. Hazard. Mater. 2011, 190, 8–27.
8. Weill, D.; Klink, G.; Roullioux, G. Plastics: The Future for Automakers and Chemical Companies. Available online: http://www.atkearney.com/paper/-/asset_publisher/dVxv4Hz2h8bS/content/plastics-the-future-for-automakers-and-chemical-companies/10192 (accessed on 12 June 2014).
9. Industry Profile for the Canadian Plastic Products Industry. Available online: http://www.ic.gc.ca/eic/site/plastics-plastiques.nsf/eng/pl01383.html (accessed on 22 July 2014).
10. American Chemistry Council. Chemistry and Light Vehicles; Economics and Statistics Department, American Chemistry Council: Washington, DC, USA, 2013.
11. European Parliament and the European Council Directive 2000/53/EC of the European Parliament and of the Council of 18 September 2000 on end of life vehicles. Off. J. Eur. Commun. I 2000, 269, 34–42.
12. Ferrao, P.; Nazareth, P.; Amaral, J. Strategies for meeting EU end of life vehicle reuse/recovery targets. J. Ind. Ecol. 2006, 10, 77–93.
13. Rebeiz, K.; Craft, A. Plastic waste management in construction: Technological and institutional issues. Resour. Conserv. Recycl. 1995, 15, 245–257.
14. Gerrard, J.; Kandlikar, M. Is European end-of-life vehicle legislation living up to expectations? Assessing the impact of the ELV directive on "Green" innovation and vehicle recovery. J. Clean. Prod. 2007, 15, 17–27.
15. Pickering, S.J. Recycling technologies for thermoset composite materials—Current status. Compos. Part A 2005, 37, 1206–1215.

16. Pimenta, S.; Pinho, S.T. Recycling carbon reinforced polymers for structural applications: Technology review and market outlook. Waste Manag. 2010, 31, 378–392.
17. Scaffaro, R.; Morreale, M.; Mirabella, F.; la Mantia, F.P. Preparation and recycling of plasticized plastic. Macromol. Mater. Eng. 2011, 141–150.
18. Brandrup, J. Ecology and economy of plastics recycling. In Proceedings of the 5th International Scientific Workshop on Biodegradable Plastic and Polymers, Stockholm, Sweden, 9–13 June 1988.
19. Conroy, A.; Halliwell, S.; Reynolds, T. Composite recycling in the construction industry. Compos. Part A 2006, 37, 1216–1222.
20. Sawyer-Beaulieu, S. Gate-to-Gate Life Cycle Inventory Assessment of North American End-of-Life Vehicle Management Processes. Ph.D. Dissertation, University of Windsor, Windsor, ON, Canada, 2009.
21. Nemry, F.; Leduc, G.; Mongelli, I.; Uihlein, A. Environmental Improvement of Passenger Cars; EC JRC-IPTS Scientific Technical Reports; European Commission: Brussels, Belgium, 2008. EUR 23038 EN.
22. Miller, L.; Sawyer-Beaulieu, S.; Tam, E. Impacts of non-traditional uses of polyurethane foam in automotive applications at end of life. SAE Int. J. Mater. Manf. 2014, 7, 711–718.
23. Kang, H.; Schoenung, J.M. Electronic waste recycling: A review of U.S. infrastructure and technology options. Resour. Conserv. Recycl. 2005, 45, 368–400.
24. Passarini, F.; Ciacci, L.; Santini, A.; Vassura, I.; Morselli, L. Auto shredder residue LCA: Implications of ASR composition evolution. J. Clean. Prod. 2012, 23, 28–36.
25. Breslin, M. Reconditioning or Recycling Plastic Auto Bumper Covers Makes Sense. American Recycler, 2010. Available online: http://www.americanrecycler.com/0110/reconditioning001.shtml (accessed on 10 June 2014).
26. Al-Salem, S.M.; Lettieri, P.; Baeyens, J. Recycling and recovery routes of plastic solid waste (PSW): A review. Waste Manag. 2009, 29, 2625–2643.
27. Sicon Technology. Polyfloat® Improves the Precision of Density Plastic Separation. Available online: http://sicontechnology.com/recycling-verfahren/plastic-recycling/polyfloat/ (accessed on 26 May 2014).
28. Shrifvars, M. Introduction to composites recycling. In Proceedings of the COMPOSITE Thematic Network Workshop, Recycling of Composite Materials in Transport, SICOMP, Pitca, Sweden, 16 June 2003.
29. Thomas, R.; Guild, F.J.; Adams, R.D. The dynamic properties of recycled thermoset composites. In Proceedings of the 8th international conference on fibre reinforced composites-FRC2000, Newcastle upon Tyne, UK, 13–15 September 2000; pp. 549–556.
30. Garcia, J.M.; Jones, G.O.; Virwani, K.; McCloskey, B.D.; Boday, D.J.; ter Huurne, G.M.; Horn, H.W.; Coady, D.J.; Bintaleb, A.M.; Alabdulrahman, A.M.S.; et al. Recyclable, strong thermosets and organogels via paraformaldehyde condensation with diamines. Science 2014, 344, 732–735.
31. BMW and Boeing. Press Release. BMW Group and Boeing to Collaborate on Carbon Fiber Recycling. 12.12.2012. Available online: https://www.press.bmwgroup.com/global/pressDetail.html?title=bmw-group-and-boeing-to-collaborate-on-carbon-fiber-recycling&outputChannelId=6&id=T0135185EN&left_menu_item=node__804 (accessed on 10 June 2014).

32. Duranceau, C.; Sawyer-Beaulieu, S. Vehicle recycling, reuse, and recovery: Material disposition from current end-of-life vehicles, SAE Technical Paper Series, 2011-01-1115. In Proceedings of the SAE World Congress, Detroit, MI, USA, 12–14 April 2011.
33. Eurostat, Statistics>Environment>Data>Database>Waste Statistics>Waste Streams> End-of-life vehicles: Reuse, recycling and recovery, Totals (env_waselvt). Available online: http://epp.eurostat.ec.europa.eu/portal/page/portal/waste/data/database# (accessed on 21 July 2014).
34. Sakai, S.; Yoshida, H.; Hiratsuka, J.; Vandecasteele, C.; Kohlmeyer, R.; Rotter, V.S.; Passarini, F.; Santini, A.; Peeler, M.; Li, J.H.; et al. An international comparative study of end-of-life vehicle (ELV) recycling systems. J. Mater. Cycles Waste Manag. 2014, 16, 1–20.
35. Sawyer-Beaulieu, S.; Stagner, J.; Tam, E. Sustainability Issues Affecting the Successful Management of End-of-Life Vehicles in Canada and the United States. In Environmental Issues in Automotive Industry—Design, Production and End-of-Life Phase; Springer-Verlag: Berlin, Germany, 2014. ISBN:978-3-642-23836-9.
36. Automotive Recyclers of Canada (ARC). A National Approach to the Environmental Management of End-of-life Vehicles in Canada: Submission to the Canadian Council of Ministers of the Environment. July 2011, p. 21. Available online: http://www.autorecyclers.ca/fileUploads/1313074351--National_ELV_EMS_approach.pdf (accessed on 29 August 2011).
37. ELVS (End of Life Vehicle Solutions Corporation). Mercury Switches, ELVS Website. Available online: http://www.elvsolutions.org/mercury_home.htm (accessed on 28 August 2011).
38. Summerhill Impact. Switch Out Program. Available online: http://www.switchout.ca/ (accessed on 31 August 2011).
39. OTS (Ontario Tire Stewardship), OTS's Rethink Tires>About Us website. Available online: http://rethinktires.ca/about-us/#sthash.HhDb08uB.dpbs (accessed on 22 July 2014).
40. Quebec (2008), Extended Producer Responsibility (EPR), Current Status, Challenges and perspec-tives, Développement durable, Environnement et Parcs. p. 145. Available online: http://www.mddep.gouv.qc.ca/matieres/valorisation/0803-REP_en.pdf (accessed on 25 August 2011).
41. Ontario 2003, Ontario Regulation 85/03 Used Oil Material, under Waste Diversion Act, 2002. Available online: http://www.e-laws.gov.on.ca/html/regs/english/elaws_regs_030085_e.htm (accessed on 22 July 2014).
42. Ontario Ministry of the Environment (OMOE). From Waste to Worth: The Role of Waste Diversion in the Green Economy, Minister's Report on the Waste Diversion Act 2002 Review; OMOE: Toronto, ON, Canada, 2009; p. 38.
43. Mark, F.; Kamprath, A. End-of-Life Vehicles Recovery and Recycling Polyurethane Seat Cushion Recycling Options Analysis, SAE Technical Paper; Society of Automotive Engineers: Warrendale, PA, USA 2004-1-0249. .
44. International Dismantling Information System. Available online: http://www.idis2.com/ (accessed on 12 May 2014).
45. Reuter, M.A.; Hudson, C.; van Schaik, A.; Heiskanen, K.; Meskers, C.; Hagelüken, C. Metal Recycling: Opportunities, Limits, Infrastructure, A Report of the Working

Group on the Global Metal Flows to the UNEP International Resource Panel; United Nations Environment Programme: Nairobi, Kenya, 2013; p. 320.
46. Dean, K.C.; Sterner, J.W.; Shirts, M.B.; Froisland, L.J. Bureau of mines research on recycling scrapped automobiles. Bur. Mines Bull. 1985, 684, 52.
47. Sawyer-Beaulieu, S.; Tam, E. Constructing a Gate-to-Gate Life Cycle Inventory (LCI) of End-Of-Life Vehicle (ELV) Dismantling and Shredding Processes; SAE Technical Paper Series 2008-1-1283; Society of Autotmotive Engineers: Warrendale, PA, USA, 2008.
48. Gesing, A.J.; Reno, D.; Grisier, R.; Dalton, R.; Wolanski, R. Nonferrous metal recovery from auto shredder residue using eddy current separators. In Proceedings of the Mineral, Metals & Materials Society Annual Meeting, EPD Congress, San Antonio, TX, USA, 16–19 February 1998; pp. 973–984.
49. Swartzbaugh, J.T.; Duvall, D.S.; Diaz, L.F.; Savage, G.M. Recycling Equipment and Technology for Municipal Solid Waste Material Recovery Facilities; Noyes Data Corporation: Park Ridge, NJ, USA, 1993; p. 150.
50. Cirko, C. Fluff afterlife—Uses for automobile shredder residue. Solid Waste Recycl. 2000, 5, 16.
51. Day, M. Demonstration of Auto Shredder Residue as a Day Cover Material for Municipal Landfills, NRCC Report, prepared for CANMET; National Research Council Canada: Ottawa, ON, 1995; ER-1352–95S. p. 58.
52. Rogoff, M.; Thompson, D.; Hilton, E. Can ADCs Work for Your Landfill? A Recent Feasibility Analysis Provided some Valuable Lessons. MSW Management, September–October 2012. Available online: http://www.mswmanagement.com/MSW/Articles/18133.aspx (accessed on 9 June 2014).
53. Mattes, A.I. Results of a Field Trial Experiment of a Hydroponic Growing System Powered By Solar Energy Using All Plastic Materials in Its Construction. The System Included Utilization of Discarded Plastic Materials Containing Heavy Metal Contaminant as a Soil Matrix; The Elevated Wetlands: Toronto, ON, Canada, 1996.
54. Takaoka, T.; Asanuma, M.; Hiroha, H.; Okada, T.; Ariyama, T.; Ueno, I.; Wakimoto, K.; Hamada, S.; Tsujita, Y. New recycling process for automobile shredder residue combined with lronmaking process. Stahl und Eisen 2003, 123, 101–106.
55. Cossua, R.; Fioreb, S.; Laia, T.; Mancinic, G.; Ruffinob, B.; Viottid, P.; Zanettib, M.C. Italian Experience on automotive shredder residue: Characterization and management. In Proceedings of the 3rd International Conference on Industrial and Hazardous Waste Management, 12–14 September 2012; p. 22.
56. Boughton, B. Evaluation of shredder residue as cement manufacturing feedstock. Resour. Conserv. Recycl. 2007, 51, 621–642.
57. Xanthos, M.; Dey, S.K.; Mitra, S.; Yilmazer, U.; Feng, C. Prototypes for building applications based on thermoplastic composites containing mixed waste plastics. Polym. Compos. 2002, 23, 153–163.
58. Lazareck, J. Two problems. One solution—Auto shredder residue/plastics composites. CIM Bull. 2004, 97, 76–77.
59. Jones, F.L. Pyrolysis of Automobile Shredder Residue; Argonne National Labaortory: Argonne, IL, USA, 1994; p. 52.
60. Day, M.; Shen, Z.; Cooney, J.D. Pyrolysis of auto shredder residue: Experiments with a laboratory screw kiln reactor. J. Anal. Appl. Pyrolysis 1999, 51, 181–200.

61. Harder, M.K.; Forton, O.T. Developments in the pyrolysis of automotive shredder residue. J. Anal. App. Pyrolysis 2007, 79, 387–394.
62. Roh, S.A.; Kim, W.H.; Yun, J.H.; Min, T.J.; Kwak, Y.H.; Seo, Y.C. Pyrolysis and gasification-melting of automobile shredder residue. J. Air Waste Manag. Assoc. 2013, 63, 1137–1147.
63. Allred, R.E.; Busselle, L.D. Tertiary recycling of automotive plastics and composites. J. Thermoplast. Compos. Mater. 2000, 13, 92–101.
64. Brown, J.D. Electrostatic plastics separation. R-net recycling technology newsletter—January. Nat. Resour. Can. 2000, 2–3.
65. Winslow, G.R.; Simon, N.L.; Duranceau, C.M.; Williams, R.L.; Wheeler, C.S.; Fisher, M.M.; Kistenmacher, A.; Vanherpe, I. Advanced Separation of Plastics From Shredder Residue; SAE Technical Paper Series 2004-1-0469; SAE: Warrendale, PA, USA, 2004.
66. Jody, B.J.; Daniels, E.J. End-of-Life Vehicle Recycling: The State of the Art of Resource Recovery from Shredder Residue; Argonne National Laboratory, Energy Systems Division: Du Page County, IL, USA, 2006. ANL/ESD/07–8.
67. Soroudi, A.; Jakubowicz, I. Recycling of bioplastics, their blends and biocomposites: A review. Eur. Polym. J. 2013, 49, 2839–2858.
68. Samacke, P.; Reed, D.B. Automotive's bioplastic future? In Proceedings of the Global Plastics Environmental Conference 2004—Plastics: Helping Grow a Greener Environment, Detroit, MI, USA, 18–19 February 2004.
69. Tseng, S.C. Using Bio-based Materials in the Automotive Industry. M.Sc. Thesis, University of Windsor, Windsor, ON, Canada, 2013.
70. Khoo, H.; Tan, R.; Chng, K. Environmental Impacts of Conventional Plastic and Bio-based Carrier Bags: Part 1: Life Cycle Production. Int. J. Life Cycle Assess. 2010, 15, 284–293.
71. Hermann, B.; Debeer, L.; Wilde, B.D.; Blok, K.; Patel, M. To compost or not to compost: Carbon and energy footprints of biodegradable biomaterials' waste management. Polym. Degrad. Stab. 2011, 96, 1159–1171.
72. Fonseca, A.; Nunes, M.I.; Matos, M.A.; Gomes, A.P. Environmental impacts of end-of-life vehicles' management: Recovery versus elimination. Int. J. Life Cycle Assess. 2013, 18, 1374–1385.
73. Stagner, J.; Sagan, B.; Tam, E. Using sieving and pretreatment to separate plastics during end-of-life vehicle recycling. Waste Manag. Res. 2013, 31, 920–924.
74. Ciacci, L.; Morselli, L.; Passarini, L.; Santini, A.; Vassura, I. A comparison among different automotive shredder residue treatment processes. Int. J. Life Cycle Assess. 2010, 15, 896–906.

PART V

OTHER METHODOLOGIES

CHAPTER 11

Case Study of a Successful Ashfill Mining Operation

TRAVIS P. WAGNER

11.1 INTRODUCTION

The latter half of the 20th century witnessed dramatic increases in personal consumption and the availability of new consumer goods fueled by unprecedented advancements in rapidly changing technology. An unintended consequence of the era of consumption has been the correspondingly dramatic increase in per capita waste generation; all of which necessitated end of life consumer management. While the implementation of the waste management hierarchy (i.e., Lansink's ladder) of reduction, reuse, and recycling have increased, globally, municipal solid waste (MSW) primarily continues to be buried as raw waste. While reliance on landfilling of MSW means that potentially valuable resources have and continue to be discarded, simultaneously we continue to mine for metals as raw materials. The mining of the global stock of metal ores increasingly is facing scarcity,

Wagner TP. "Case Study of a Successful Landfill Mining Operation." SUM 2014, Second Symposium on Urban Mining Bergamo, Italy; 19 – 21 May 2014. © CISA Publisher (2014). Used with permission from Eurowaste Srl and the publisher.

geopolitical challenges, and concerns over the associated environmental impacts. Consequently, there is recognition that vast amounts of comparatively concentrated, valuable materials reside in relatively shallow surface deposits in current and former landfills that are relatively close to industrial centers. This is especially the case with older landfills where source separation of recyclable materials rarely occurred (Kaartinen, Sormunen, and Rintala, 2013; Quaghebeur et al., 2013). The discarding of high-value materials while simultaneously mining for these same materials is a highly unsustainable practice.

As global demand and competition for limited resources increases, costs of raw materials will increase. This has resulted in increased attention to the potential to mine landfills (Krook, Svensson, and Eklund, 2012). According to Savage et al. (1993, p. 58), "Landfill mining involves the excavation of completed fill to reclaim resources and in doing so, reclaim capacity." From a technological perspective, landfills can be mined. But, as with any mining operation, the primary decision factor is economics—will the economic value of the mined material exceed the cost of mining, segregating, processing, transporting, and other operational factors? With landfill mining, there are other potential economic benefits, which can include value of additional landfill capacity, recovery of energy values, and avoided remediation costs. Another crucial consideration different than traditional mining is understanding how a mining operation could affect an operating or closed landfill with regards to leachate generation, groundwater contamination, contaminated surface water run-off, odors, and air emissions.

Mining landfills is not a new concept (Jones et al., 2013). Historically the focus of landfill mining has been to extend the life of landfills by increasing landfill space, to conduct remediation on problem landfills, to reclaim land, and to extract methane gas (Krook et al, 2012). According to the Florida Department of Environmental Protection (2009), as of 2009, there were 32 current or former landfill mining projects in the US. The objectives of landfill mining at these sites was to remediate groundwater contamination at unlined landfills, creating additional airspace and thus increasing landfill capacity and life, reducing the cost of closure, recovery of metals, and recovering energy by using landfill materials as refuse derived fuel (Florida Department of Environmental Protection, 2009). However, the most common objective of landfill mining in the US has been to relocate previously disposed of waste from an unlined landfill unit to an adjacent lined unit without processing or recovery of materials (Florida Department of Environmental Protection, 2009). Krook, Svensson, and Eklund (Krook et al, 2012) observed that in the US, interest in landfill mining peaked during the 1990s primarily in response to then strengthening national, minimum requirements for MSW landfills, which included requirements for liners, groundwater monitoring, and corrective action.

Research on landfill mining continues to focus on its feasibility and on pilot studies, especially related to the characterization of landfilled materials (Krook et al, 2012). The potential for landfill mining is significant. In the EU alone there are an estimated 150,000 to 500,000 closed and active landfills with an average size of 8,000 m^2 (Krook et al, 2012), which collectively contain some 30–50 billion m^3 of waste (Ratcliffe, Prent, and van Vossen, 2012). In the US, 6,270 MSW landfills closed between 1988 and 2005 (U.S. Environmental Protection Agency, 2013) in addition to numerous landfills and dumps having closed previously; as of 2011, there were 1,908 operating MSW landfills in the US (U.S. Environmental Protection Agency, 2013).

When mining for recoverable materials, the fundamental steps in landfill mining are similar to traditional mining: excavation and processing (e.g., beneficiation, crushing/grinding, sizing, and concentrating) materials for intended market—secondary smelting for metals. The mining of heterogeneous wastes that are not completely decomposed creates unique technological challenges and localized problems such as odor and vector nuisances. Landfilled heterogeneous waste is basically composed of organics, minerals, metals, and water (U.S. Environmental Protection Agency, 2013). However, especially for older landfills, the lack of knowledge and recordkeeping regarding the specific wastes disposed of presents additional challenges thus raising the costs of sampling and analyses to properly characterize the waste. With regards to landfilling mining, metals generally are the easiest materials to recover and have the highest value (Ford et al., 2013). Based on research by Quaghebeur et al. (2013), the concentration of metals in waste excavated from an MSW landfill, which ranged from 3% to 6% w/w, was comparable to the concentration of the originally landfilled waste. The authors found that the metals fractions consisted mainly of ferrous metals and aluminum comprised the majority of the non-ferrous fraction (Quaghebeur et al., 2013). According to Ratcliffe et al. (2012), based on results from 60 landfill mining projects in the literature, the mean concentration of total metals in landfilled waste was 2.0%. In the US, on average, discarded MSW containes 4.98% metal. In a 2011 study, the metal content of discarded residential MSW in the US state of Maine was 3.3% (Criner and Blackmer, 2012), which is likely lower than the national average because Maine has a beverage container deposit/refund program that includes aluminum cans and the characterization was limited to residential, bagged waste.

An integrated approach to maximizing the recovery of value and material from heterogeneous waste mined from a landfill is enhanced landfill mining (ELFM). As defined by Jones et al. (2013, p. 48), ELFM is "the safe conditioning, excavation and integrated valorization of (historic and/or future) landfilled waste streams as both materials…and energy…, using innovative transformation technologies and respecting the most stringent social and ecological criteria." That is, mined waste is processed through multiple means to maximize the recovery of energy and material value of the waste in a cost-effective manner. For example, landfilled heterogeneous waste can first be burned as refuse derived fuel to recover the energy value while simultaneously reducing the volume of the waste by up to 90% transforming into an ash. Because of the dramatically reduced volume, metals become concentrated and the ash has a smaller and more uniform size. The concentrated and smaller particle sizes allow for more efficient and cost-effective sorting and processing to recover metals.

By extension, ash from WTE facilities, which has already recovered the raw waste's energy value, has been "pretreated" by eliminating the organic, water, and low value portions of raw waste. The result is that any metals have been concentrated and a more uniform, smaller particle sized-waste has been produced. Consequently, the ash is more amenable to dry sorting and processing to recover metals.

The specific chemical and physical characterization of ash from WTEs depends on the characteristics and components of the raw MSW and the operational conditions of the WTE. MSW incinerator ash is comprised of both bottom ash and fly ash; however, the majority of ash is from bottom ash and contains slag, glass, ceramics, uncombusted materials, and incompletely combusted organics (Williams, 2005) and contains significant concentrations of ferrous and non-ferrous metals, including Al, Ag, Cu, Pb, Sn, and Zn, which are often found as small pieces (Reijnders, 2005). The management of ash in the US has been primarily as co-disposal with raw MSW in landfills, used as alternative daily cover, or in some cases disposed of in dedicated ash monofills in which only ash is accepted for disposal (Cardoso, Levine, and Rhea, 2008). The US currently has 85 WTEs located in 22 different states (U.S. Environmental Protection Agency, 2013). In 2011, WTEs in the US accounted for 21.8% of post-recovery disposal compared to 78.2% for landfilling (U.S. Environmental Protection Agency, 2013). According to Reijnders (2005), in 2000, approximately 25 Mt/year of MSW fly ash was generated in the USA, EU, and Japan. In Europe, there are 472 WTEs in 18 countries (International Solid Waste Association. (2012). Although the number of WTEs in the US is decreasing, the number of WTE plants in Europe is increasing. In 2009, 16 million Mt of bottom ash was produced from 449 European WTEs (CWEEP, 2011).

Given the history of disposing of large amounts of valuable materials, especially metals, in thousands of landfills, an intriguing question arises; can landfills be successfully mined specifically for recovering material resources without government subsidies? Given the spotty history of successful landfill mining in the US, the more specific question is can ash monofills be successfully mined for metals? This paper is a case study that examines the first successful ash monofill mining operation for ferrous and non-ferrous metals in North America at the ecomaine ash monofill in the US state of Maine.

11.2 CASE STUDY INTRODUCTION

The facility examined in this case study is owned by ecomaine, a nonprofit regional waste disposal and recycling operation owned and operated by 21 municipalities. ecomaine, which is located in the southern part of the US state of Maine (see Figure 1), handles MSW from approximately 25% of the state's residential population of 1.32 million. ecomaine has operated a mass burn WTE since 1988. The WTE is licensed to process 499 Mt of

MSW per day and generates about 96,000 megawatts of electricity per year. ecomaine also owns and operates a 101-hectare landfill/ash monofill, which is located 5 km from the WTE. The ash monofill portion is 8 ha. From 1978 until 1988, the landfill accepted raw, baled MSW. Starting in 1988, with the start-up of the WTE, only ash has been landfilled. In 1990, a new recycling facility was built, which processed and diverted 5,945 kg from the wastetsream in the first year. In 2013, the recycling facility collected and diverted 15,932 kg representing a 168% increase in diversion. The state of Maine has a beverage container deposit/refund system, which includes aluminum containers, with an estimated 80% capture rate. The ash monofill held an estimated 725,700 Mt of ash, which included unknown, but significant amounts of ferrous and non-ferrous metals.

On the tipping floor, MSW is transferred by escalator into the fire. While on the tipping floor, no materials (e.g., metals) are removed. As is typical with WTEs, the volume of the waste is reduced by about 90%. The resulting ash (bottom and fly ash) is conveyed where it passes under a ferrous and rare earth magnet and then to trucks where it is transported to the nearby ash monofill. Between 1988 and 2004, the ash included post-burn metal at a concentration of 10.5% ferrous. In 2004, a magnetic separator was installed at the end of the WTE that was designed to remove post-burn ferrous metals from the ash, which resulted in a mean ferrous content of 6% in the ash when landfilled. Also in 2004, Maine banned the disposal of electronic waste, which has reduced the amount of metal in MSW. In June 2012, a rare-earth magnet was installed at the WTE, which dropped the concentration of ferrous metal even lower. Post-burn metal is now collected and sold to a secondary smelting market.

WTE staff to excavated, sampled, and analyzed ashfill materials to determine the feasibility of recovering ferrous metals in the ash monofill. Staff estimated that recovery of metals from the ecomaine ash monofill would be successful based on preliminary estimates of a ferrous metal concentration in the ash. Only the ash buried from 1988 until 2004 was targeted for potential mining because 2004 was when the first post-burn magnetic separator was installed.

A major concern of the project was the need to expedite the mining operation because active mining, specifically excavation, will increase the potential for leachate generation, especially in a moderately wet climate (1100 mm/year) like Maine (National Weather Service, 2013). The generation of leachate is of special concern because there is no liner under the pre-2004 ash monofill portion. Consequently, a monthly-leasing arrangement was established to impose an economic incentive to expedite the mining process and to discourage the stockpiling of mined metals onsite in an effort to maximize market returns from price speculation.

11.2.1 MINING OPERATIONS

The mining operation began in November 2011 with the first shipment of mined metals occurring in December 2011. All processing is done onsite. The first step is excavation of the existing ash. Excavation occurs with a frontend loader. Excavated material is moved from the pit to a temporary stockpile near the processing equipment. As shown in Figure 2, excavated ash is processed through dry separation and then screened to separate finer fractions and metals from larger materials using a series of conveyers and shaker screens. Material that does not pass the 50 mm screen is temporarily stored and sent back to the ashfill where it is compacted, graded, and covered with a temporary geomembane until the mining operation is completed. Material is shredded and crushed then goes to the overband magnet and then to a hand sorting table and the ferrous (mostly steel and iron) and non-ferrous (mostly aluminum and stainless steel) metals are captured. The operation has processed approximately 450 Mt of ash per day. Recovered ferrous and non-ferrous metals are contaminated with ash (about 14% by weight), by agreement the ash component can be returned back to the ash monofill.

11.2.2 MINING RESULTS

Between November 2011 and November 2013, 199,580 Mt of ash had been processed. It is important to note, however, some of this ash has been processed more than once. The original focus of the mining operation was ferrous metals. However, potentially recoverable quantities of non-ferrous metals (aluminum and stainless steel) were observed. An eddy current separator was subsequently installed and the ash previously processed for ferrous metals was reprocessed to recover non-ferrous metals. Since the installation of the eddy current separator, all mined ash is now subjected to both the magnet and eddy current separator. During the operation period, 19,958 Mt of metals were segregated and recovered and sent offsite for processing and secondary smelting. The metals collected include nails, steel cans, aluminum containers, automotive parts, springs from mattresses, plumbing parts, and miscellaneous metal items. The recovered metal shipped offsite requires extensive processing including shredding and grading to separate different grades of material (e.g., steel, cast iron, sheet metal etc.). Although the percentages of the different types of items is not known, the metal shipped offsite is categorized as large (>127 mm), medium (between 50 mm and 127 mm), small (<50 mm), and non-ferrous zorba. Non-ferrous metals constituted 5% of the recovered metal. Based on the October 2013 price paid for post-burn metal in Maine, which was US$94.97 (€69.88) per Mt, the estimated scrap value of the recovered metal was US$1,895,520 (€1,394,665). This however is a very conservative estimate because it assumes all metals were ferrous, which is not accurate as 5% are zorba—mixed non-ferrous metals, so any revenues greater than €69.88 per Mt received for non-ferrous metals are not included. The landfill owner, ecomaine, was paid US$529,000 (€390,838) for the value of the recovered metal. In addition, 8,028 m^3 of material was removed thereby increasing the airspace of the ash fill. The value of this additional space as determined by landfill staff is US$267,000 (€361,300).

The capital costs and operational costs for this landfill mining project are not available because these costs are proprietary. Clearly the costs are below the resale value of the recovered metals, which is shared with ecomaine in addition to leasing costs of the site. According to Hogland et al. (2011), the per tonne capital costs for landfill mining in the US ranges from US$10 to US$30 (€7.30-€22.09) and per tonne operational costs range from US$30 to US for a total cost of US$40-US$120 per tonne. Ratcliffe et al. (2012) identified per tonne costs for separating landfill mined waste to be €45. A Scottish study (Ford et al., 2013) assumed a per tonne capital cost for excavation and separation to be €6 to €12 and the per-tonne operational cost for excavation and separation to be €30.10 to €60.25. However, per tonne costs at the ecomaine

FIGURE 1: Location of the ashfill mining operation.

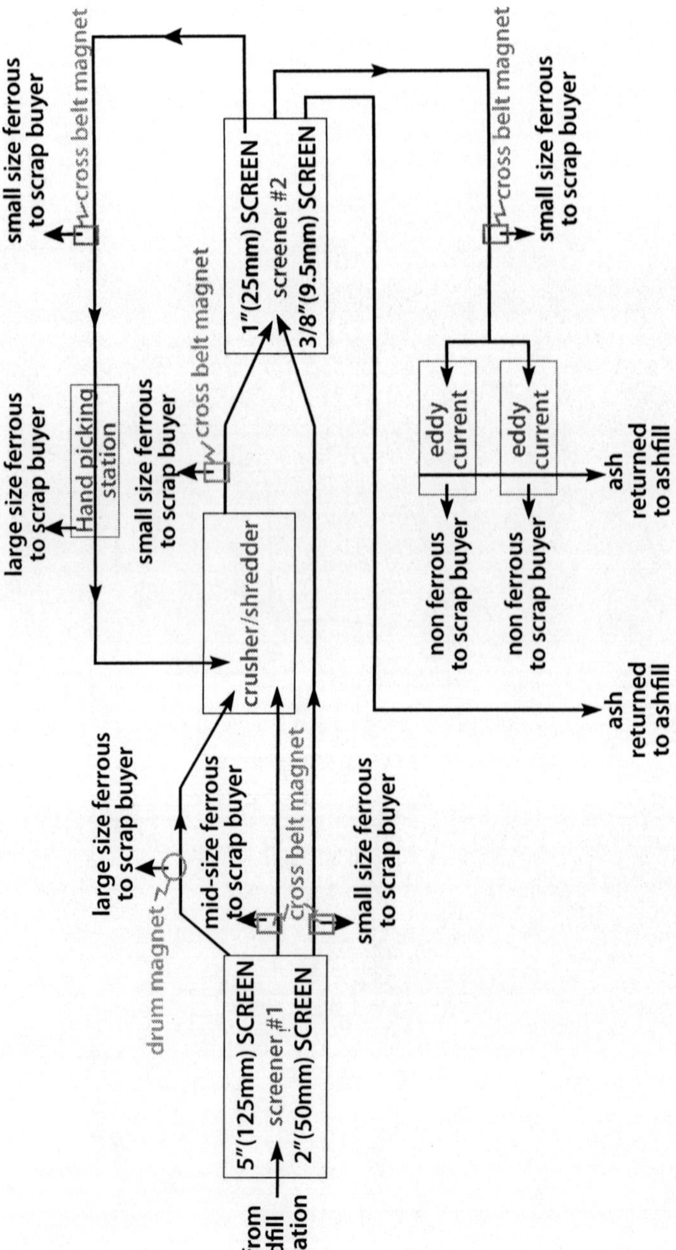

FIGURE 2: Schematic of the landfill mining operation.

ash monofill are presumed to be far lower because of the lower segregating and processing costs of the ash with its comparatively concentrated metals.

11.3 CONCLUSION

As society continues to be characterized as a disposable society, MSW landfills present an attractive opportunity to recover materials, but they also present a technological and economic challenge because of the undesirable characteristics and contents of MSW from a processing perspective. MSW landfills contain heterogeneous waste, which is comprised of organics in various stages of decomposition in addition to hazardous wastes, contained and uncontained gases, bulky items, and valueless inert materials. In contrast to MSW landfills, landfills dedicated solely for the disposal of ash (ash monofills) from the combustion of MSW are more amenable to cost-effective mining. The waste (i.e., the ash) has already been "pre-processed" by concentrating the metals and eliminating much of the undesirable portions MSW (e.g., organics, hazardous waste, and bulky items). The result is analogous to mining for metal ore.

This case study documented successful mining operation, which is mining 20 years worth of MSW incinerator ash. Because this is a novel project, there are challenges, opportunities, and unexpected outcomes. The result is lessons learned that contributes to the body of knowledge for MSW landfill mining. Illustrative of the lessons learned aspect of the mining project is the reprocessing of previously processed ash to recover non-ferrous metals highlighting the learning curve, adaptability, and investment needs (the eddy current separator) of a successful mining operation. A small positive unexpected consequence has been the increased airspace of the ash monofill.

While this case study has shown that ash monofill mining can be successful, ash monofills represent a very small percent of the universe of landfills. There is a much bigger challenge with raw waste and codisposal, which comparatively have a much lower metals concentration and greater processing challenges. The emergence of ELFM may be able to overcome the challenges of mining raw waste as it seeks to extract energy value and concentrate the metals, which can be removed through post-burn magnets and eddy current separators similar to the ecomaine ash monofill. This case study demonstrates that mining of combusted MSW specifically for the recovery of metals can be successful.

REFERENCES

1. Cardoso A.J., Levine A.D. and Rhea, L.R. (2008). Batch test assessment of waste-to-energy combustion residues impacts on precipitate formation in landfill leachate collection systems. J Air Waste Manage, 58:19–26.
2. Confederation of European Waste-to-Energy Plants [CWEEP]. (2011). Environmentally sound use of bottom ash. Available from www.cewep.eu/information/publicationsandstudies/statements/ceweppublications/m_722.
3. Criner G.K. and Blackmer T.L. (2012). 2011 Maine Residential Waste Characterization Study. University of Maine, School of Economics Staff Paper #601. Available from http://umaine.edu/wcs/publications-and-reports.
4. Florida Department of Environmental Protection [FDEP]. (2009). Landfill reclamation demonstration project, Perdido Landfill. Escambia County Neighborhood and Community Services Bureau Division of Solid Waste Management. Prepared by Innovative Waste Consulting Services, LLC. Available from http://www.dep.state.fl.us.
5. Ford S., Warren K., Lorton C., Smithers R., Read R., and Hudgins M. (2013). Feasibility and viability of landfill mining and reclamation in Scotland. Available from www.zerowastescotland.org.uk/content/report-feasibility-viability-landfill-mining-and-reclamation-scotland.
6. Hogland, W., Hogland, M., and Marques, M. (2011) Material Recovery, Energy Utilization and Economics of Landfill Mining in the EU (Directive) Perspective. Presented at the Sustainable Solid Waste Management in Eastern Europe - Prospects for the Future, 19-20 May, Ukraine.
7. International Solid Waste Association. (2012). Waste-to-energy state-of-the-art-report, 6th Edition, (2012). Available from www.cewep.eu/information/data/iswawtestateoftheartreport/index.html.
8. Jones T.P., Geysen D., Tielemans Y., Van Passel S., Pontikes Y., Blanpain B., Quaghebeur M., and Hoekstra, N. (2013). Enhanced Landfill Mining in view of multiple resource recovery: A critical review. J Clean Prod, 55: 45-55.
9. Kaartinen T., Sormunen K. and Rintala J. (2013). Case study on sampling, processing and characterization of landfilled municipal solid waste in the view of landfill mining. J Clean Prod, 55: 56-66.
10. Krook J., Svensson N. and Eklund M. (2012). Landfill mining: A critical review of two decades of research. Waste Manage, 32: 513–520.
11. National Weather Service. (2013). U.S. National Oceanic and Atmospheric Administration. Available from www.erh.noaa.gov/er/gyx/climate_f6.shtml.

12. Quaghebeur M., Laenen B., Geysen D., Nielsen P., Pontikes Y., Van Gerven T., and Spooren J. (2013). Characterization of landfilled materials: screening of the enhanced landfill mining potential. J Clean Prod, 55: 72-83.
13. Ratcliffe A., Prent O.J., and van Vossen W. (2012). Feasibility of material recovery from landfills (MFL) in the European Union. The ISWA World Waste Congress 2012, Florence, Italy, 17-19 September, 2012.
14. Reijnders L. (2005). Disposal, uses and treatments of combustion ashes: a review. Resour Conserv Recy, 43:313–336.
15. Savage G.M., Gouleke C.G., and Stein, E.L. (1993). Landfill mining: Past and present. BioCycle, 34: 58-61.
16. U.S. Environmental Protection Agency. (2013). Municipal Solid Waste (MSW) in the United States: 2011 Facts and Figures. Office of Solid Waste, Washington, DC. Available from www.epa.gov/epawaste/nonhaz/municipal/msw99.htm.
17. Williams P.T. (2005). Waste Treatment and Disposal (2nd ed.). Chichester, UK, John Wiley and Sons.

CHAPTER 12

Multi-Stage Control of Waste Heat Recovery from High Temperature Slags Based on Time Temperature Transformation Curves

YONGQI SUN, ZUOTAI ZHANG, LILI LIU, AND XIDONG WANG

12.1 INTRODUCTION

It is well known that the steel industry is energy intensive, consuming around 9% of anthropogenic energy [1] and emitting large quantities of CO_2 into the atmosphere [2]. With the acceleration of global warming nowadays, energy saving and CO_2 emission reduction in the steel industry is attracting more and more attention, although the energy efficiency has already been substantially improved by implementing extensive advanced technologies. According to the previous estimations [3,4], high temperature (1450–1550 °C) slags, carrying a substantial amount of high quality heat, represent the last potential source for energy reduction in the steel industry. In China, the steel industry's output of crude steel was more than

710 million tons in 2012 [5], and accordingly around 200 million tons of high temperature blast furnace slags (BF slags) and 70 million tons of steel slags were produced [6] and the total waste heat was more than 4.80 $\times 10^{19}$ J, equivalent to 16 million tons of standard coal, whereas less than 2% of that was recovered, according to the estimation of Cai et al. [7], so there is a great potential of waste heat recovery. BF slags alone accounted for more than 70% of the waste heat of the slags in the steel industry, and therefore the previous studies were mainly focused on BF slags [8].

Conventionally, BF slags can be treated through two methods: gradually cooled by air in a slag pit [9] or rapidly quenched by water [10]. Air-gradually-cooled slags have low utilization values because of their weak hydraulicity due to the high content of crystalline phases [11], while water quenched BF slags are increasingly utilized as cementitious materials because of the glassy phases [11] and the similarity between the components of slags and Portland cement, high content of calcium silicates [12]. However, the water quenched method faces a series of problems, such as water consumption and pollution, SO_2 and H_2S emissions and energy consumption and waste [13,14]. To solve the aforementioned problems and realize the recovery of waste heat, a dry granulation method has been proposed, through which the liquid slags are granulated into small droplets with the diameter of several millimeters using different waterless granulation technologies, such as rotary cup atomizer [15–18], rotating drum [19,20], air blasts [21,22] and so on. Recently the combination of dry granulation with other waste heat utilization methods, such as hydrogen production from biogas [18,23], coal gasification [24,25] and heat storage of phase change materials (PCM) [26] has been intensively studied, and is expected to be a promising method in the future. Whatever the utilization method applied, the understanding of the variation of slag properties during cooling processes and the control of heat transfer are fundamental for simultaneously realizing waste heat recovery and slag recycling. Considering the large amount of BF slags, the present study was focused on BF slags. Meanwhile, iron ore has been degraded and Al_2O_3 content in gangues has been increasing in the past decades [27], which has led to variations in the chemical composition of BF slags, such as an increase of Al_2O_3 content. With the variation of chemical compositions of slags, the slag properties, including crystallization properties could be changed, which could influence the waste heat recovery of high temperature slags. Three samples containing different levels of Al_2O_3 were therefore designed in this study.

The investigation was carried out using a Single Hot Thermocouple Technique (SHTT) for visualizing phase changes in the slag melts, through which Time Temperature Transformation (TTT) curves were established. TTT curves described well the variation of slag properties, based on which a multi-stage control method was proposed in this paper. The potential of waste heat recovery at different stages was accordingly calculated.

12.2 MATERIALS AND METHODS

12.2.1 SAMPLE PREPARATION

In the present study, three slag samples with a CaO/SiO_2 ratio of 1.05 and Al_2O_3 contents of 10–25 wt% were prepared using analytically regent (AR) pure oxides. These oxides were mixed and pre-melted in a molybdenum (Mo) crucible (Φ40 mm × 45 mm × H40 mm) under an argon atmosphere at 1500 °C for 2 h to homogenize the chemical compositions. Then the liquid slags were rapidly poured into the cold water to obtain glassy phases. Subsequently, the solid slags were dried at 120 °C for 12 h, crushed and ground to 300 mesh size for compositional analysis and SHTT experiments. The chemical compositions were measured by the X-Ray fluoroscopy (XRF) technique and the results are listed in Table 1. To confirm the glassy phases of slags and the accuracy of SHTT experiments, X-ray diffraction (XRD) tests were carried out and the results are shown in Figure 1.

TABLE 1: Chemical compositions of samples (wt%)

Samples	Basicity (B)	CaO	SiO_2	MgO	Al_2O_3
A1(XRF)	1.05	40.7%	38.9%	9.2%	11.3%
A2(XRF)	1.02	35.8%	35.1%	9.1%	20.0%
A3(XRF)	1.02	34.3%	33.5%	9.0%	23.1%

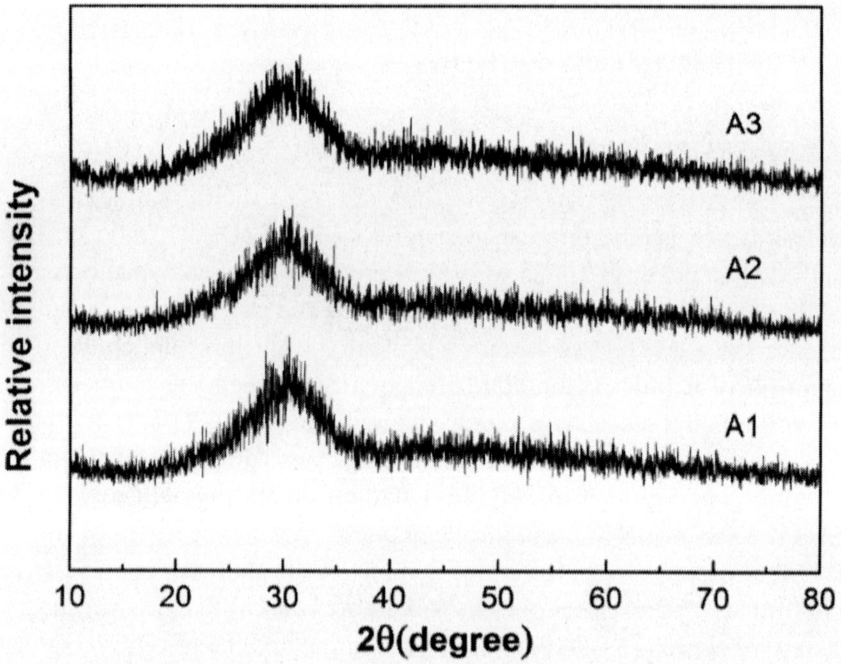

FIGURE 1: XRD results of the pre-melted slags.

12.2.2 APPARATUS AND PROCEDURE

Experiments in this study were carried out using SHTT for visualizing the phase transformations and measuring the incubation time of these slags. SHTT combines the advantages of in-situ optical observation and the low inertia of the system, with a maximum cooling rate of 200 °C/s. The work mechanism of SHTT has been described in detail elsewhere [28,29] and is only briefly outlined here. As schematically shown in Figure 2, a Pt-Rh thermocouple was used to heat the sample and simultaneously measure the temperature, which was controlled by a computer program. A microscope equipped with a video camera was used to observe and capture images of the slag melts. The isothermal experiments by SHTT were performed following several steps as below. Firstly, the temperature of thermocouple was calibrated using pure K_2SO_4 with a constant melting point 1067 °C. Secondly, about 10 mg sample was mounted on the top of the thermocouple, heated to 1500 °C, and held for 120 s to eliminate the bubbles and homogenize the chemical composition. Thirdly, the liquid slags were rapidly quenched at a cooling rate of 50 °C/s to a given temperature and held for a long time at this temperature. Then the evolution of crystallization with time in the slag melts was observed and the sample images were captured by the video camera, shown in Figure 3 as an example. The crystalline phases precipitated in the melts were identified by the XRD technique.

After the liquid slag was rapidly quenched to a given temperature, it may take some time to crystallize, i.e., there is a certain incubation time. To reduce measurement errors, the incubation time at each temperature point was measured at least three times and the average value was used. The obtained incubation time can help to design a waste heat recovery process in a reasonable time to avoid crystal formation. The variation of incubation time with isothermal experiments was presented in TTT curves. The crystalline phases precipitated in slags were confirmed by Factsage software calculation and the latent heat of phase transformation was calculated [30].

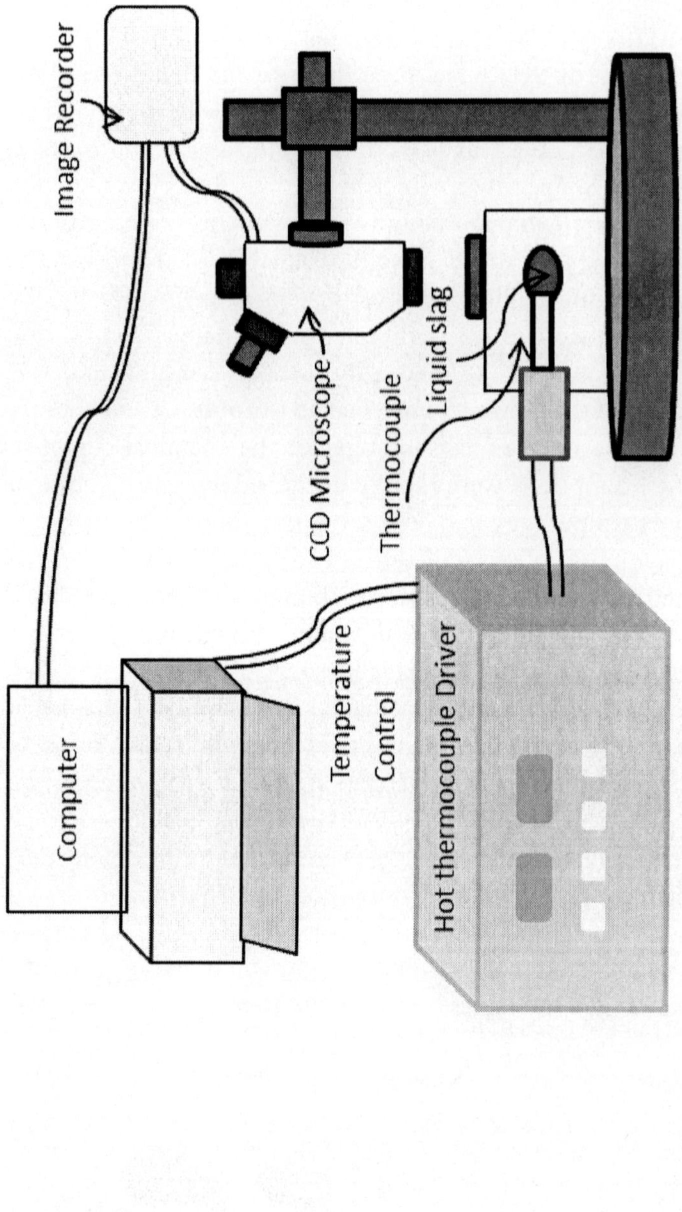

FIGURE 2: Schematic diagram of the SHTT instrument.

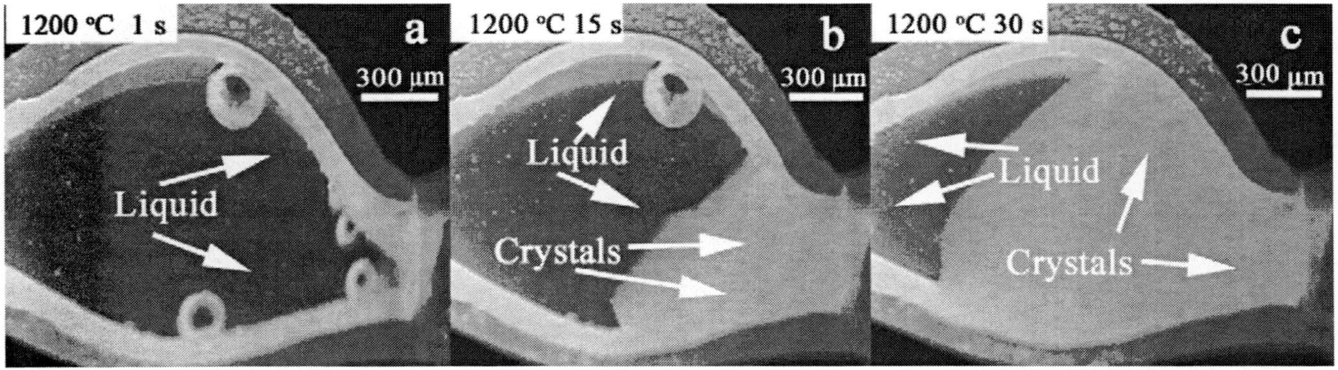

FIGURE 3: Schematic diagrams of SHTT images for sample A1 at 1200 °C.

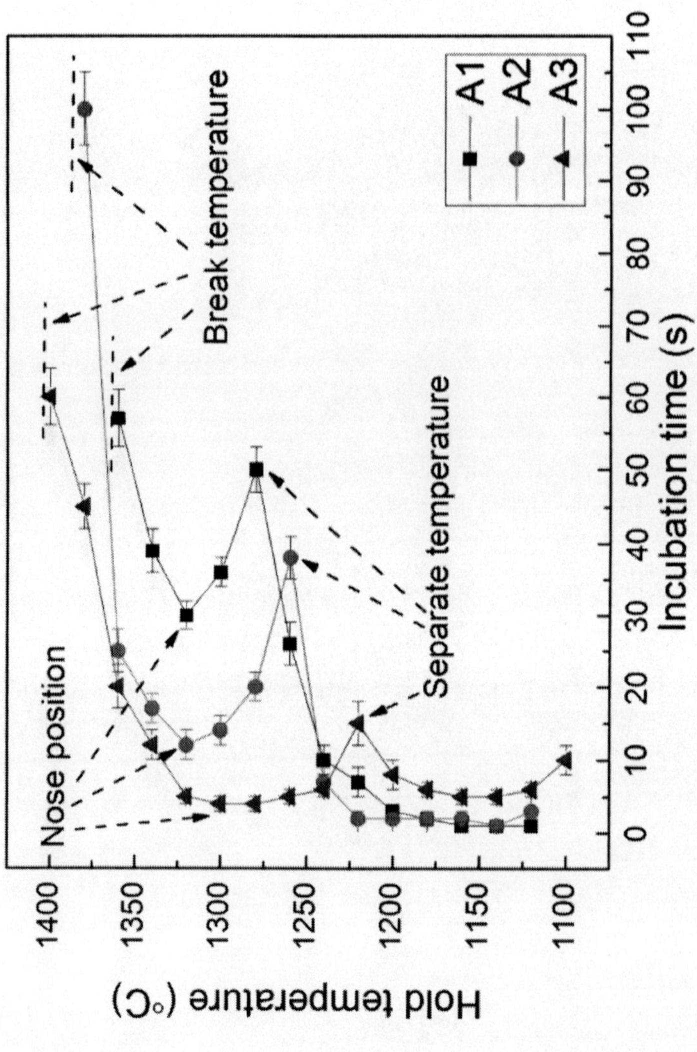

FIGURE 4: TTT curves of the different samples.

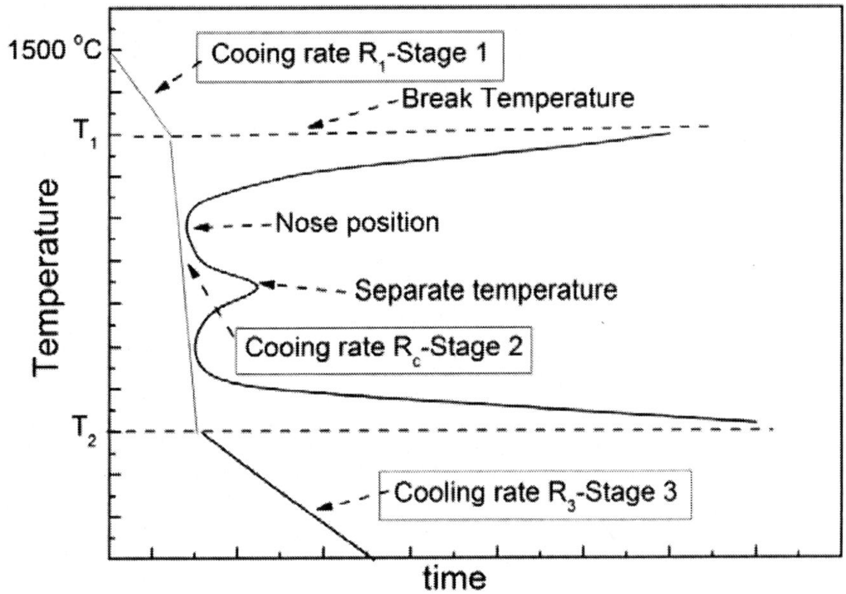

FIGURE 5: Schematic diagram of the multi-stage control of waste heat recovery.

12.3 RESULTS AND DISCUSSION

12.3.1 MULTI-STAGE CONTROL OF WASTE HEAT RECOVERY BASED ON TTT CURVES

12.3.1.1 TTT CURVES

The TTT curves of different samples were presented in Figure 4. As can be seen, these TTT curves showed a similar shape, that is a double "C" from high temperature to low temperature, which suggested that two different crystallization events occurred. As the slag melt was quenched from 1500 °C, the crystallization was not observed when it was higher than the break temperature (the highest temperature at which crystallization could appear), indicating that the waste heat recovery from the slags can last a long time during this temperature interval. It can be seen that the break temperature increased from sample A1 to A3, which suggested that the crystallization was enhanced by Al_2O_3 addition.

As the temperature decreased lower than the break temperature, crystals may precipitate after an incubation time. The incubation time decreased with the decrease of isothermal temperature until the nose position, then it increased with further decreasing temperature to the separate temperature between these two "C" shapes. With the decrease of temperature, the undercooling degree was increased, which was beneficial for crystal formation, whereas the viscosity increased at the same time, which suppressed the crystal formation. These two factors caused by temperature decrease influenced the crystal precipitation in opposite directions and caused the appearance of a shortest incubation time, i.e., at the nose position. The shortest incubation time at the nose position suggested that the waste heat recovery time with medium must be very quick at this temperature in order to avoid crystallization.

With further decreasing temperature, a similar variation tendency as with the first "C" shape was observed in the crystallization temperature range. Finally the temperature was decreased to a low temperature zone where no crystallization was observed due to the high resistance of mass transfer caused by the higher viscosity. These results indicated that the waste heat recovery time from slags could be set up with temperature, and the cooling process was therefore determined based on the TTT curves. An apparent trend was observed in that the TTT curves moved to the left side in high temperature zones with increasing Al_2O_3 content, which indicated that the crystallization was enhanced by Al_2O_3. This suggested that an increasing Al_2O_3 content resulted in a lower waste heat recovery time in this temperature interval. It is also noted that the second "C" shape in low temperature zones did not show an apparent variation tendency for different samples, which might be because that the effect of undercooling on the crystallization was greater than that of the Al_2O_3 content [31].

12.3.1.2 MULTI-STAGE CONTROL METHOD

Based on the aforementioned findings, a multi-stage control method of waste heat recovery from high temperature slags was proposed here, a general schematic diagram of which is shown in Figure 5. During stage 1, as the temperature was higher than the break temperature, no crystallization was observed and the heat exchange could continue for a long time to fully extract the waste heat. Slags in this temperature zone could be slowly cooled and the high quality waste heat could be fully exchanged to PCM [26] and stored for further utilization. This stage was located in the temperature range where slags were liquid, defined as Liquid region. During stage 2, with the temperature decreasing to the first "C" and the second "C" shapes, the heat exchange time should be adjusted according to the incubation time. The cooling rate during this stage must be larger than the critical cooling rate, which was a cooling rate larger than which crystallization does not appear. The critical cooling rate can be calculated by the following expression [32]:

$$R_c = \frac{1500 - T_0}{\tau} \quad (1)$$

where R_c is the critical cooling rate; T_0 and τ are the crystallization temperature and incubation time of nose position. The calculated values of R_c

were 6 °C/s, 15 °C/s and 50 °C/s for samples A1, A2 and A3, respectively. As can be seen, the variation of Al_2O_3 content greatly influenced the critical cooling rate and these increasing critical cooling rates could bring new challenges for waste heat recovery of high temperature slags, which must be considered in any practical waste heat recovery process. In this stage, the liquid slags should be rapidly quenched at a cooling rate larger than critical cooling rate to avoid crystal formation. This stage was located in the crystallization temperature range, defined as Crystallization region. During stage 3, as the temperature further decreased, crystallization was not found in the slag melt due to the high transfer resistance of ions, i.e., the slag started to solidify under this temperature, and therefore there was enough time to extract the waste heat from the slags. This stage was suitable for combination of waste heat recovery from high temperature slags with other new heat utilization methods, such as H_2 production, coal gasification and heat storage in PCM because the slags have been totally solidified. Purwanto et al. [23] have studied the hydrogen production using biogas and hot slags and the temperature range in their study was from 973 K to 1273 K, which was located in the temperature zone of solidified slags. They found that the slag acted not only as a good thermal media but also as a good catalyst for H_2 production. The possibility of CO_2/coal gasification has been studied by Li et al. [24,25] using slag granules as heat carries for waste heat recovery from BF slags in the temperature range of 1223 and 1423 K and they found that the added BF slags greatly enhanced the gasification reaction, which might be a development trend for waste heat recovery form high temperature slags in the future. This stage was located in the temperature range where only solidified slags existed, defined as the Solid region. However, it should be pointed out that attention should be paid to the recrystallization from the solidified glassy slags during this stage. Recrystallization caused the increase of the content of crystalline phases in slags and finally reduced the utilization value of solid slags.

In this section, the whole temperature range during waste heat recovery process was divided into three regions, i.e., Liquid region, Crystallization region and Solid region, and the technological parameters of heat exchange can be therefore designed according to the TTT curves, as shown in Figure 5.

12.3.2 FITTING OF TTT CURVES

To realize a continuous control of waste heat recovery from high temperature slags, it is necessary to obtain a fitting of the TTT curves for a practical waste heat recovery process. The polynomial expressions could be used to fit the TTT curves, from which the incubation time at a given temperature could be deduced [33]. The following equation gave acceptable calculation results that fitted the measured incubation time for samples in this study:

For sample A1:

$$\tau = 31146.8458T - 35.41T^2 + 0.01789T^3 - 3.38 \times 10^{-6}T^4$$

$$1280^{\circ}C \leq T \leq 1360^{\circ}C,\ R^2 = 0.999 \qquad (2.1)$$

$$\tau = 546787.44 - 1871.79T - 2.40T^2 + 0.09137T^3 + 2.93 \times 10^{-7}T^4$$

$$1120^{\circ}C \leq T \leq 1280^{\circ}C,\ R^2 = 0.994 \qquad (2.2)$$

For sample A2:

$$\tau = 1.41 \times 10^7 - 43010.47T + 49.25T^2 - 0.02506T^3 + 4.78 \times 10^{-6}T^4$$

$$1260^{\circ}C \leq T \leq 1380^{\circ}C,\ R^2 = 0.975 \qquad (3.1)$$

$$\tau = 2.34 \times 10^6 - 7964.41T + 10.16T^2 - 0.00576T^3 + 1.23 \times 10^{-6}T^4$$

$$1120^{o}C \leq T \leq 1260^{o}C,\ R^2 = 0.987 \quad (3.2)$$

For sample A3:

$$\tau = 11178.25 - 47.15T + 0.0726T^2 - 4.85 \times 10^{-5}T^3 + 1.18 \times 10^{-8}T^4$$

$$1220^{o}C \leq T \leq 1400^{o}C,\ R^2 = 0.970 \quad (4.1)$$

$$\tau = 761414.07 - 2633.75T + 3.42T^2 - 0.00197T^3 + 4.26 \times 10^{-7}T^4$$

$$1110^{o}C \leq T \leq 1220^{o}C,\ R^2 = 0.997 \quad (4.2)$$

where τ was incubation time and T was the holding temperature. These calculated polynomial expressions well matched the TTT curves, which could provide significant information for tailoring the technological parameters for waste heat recovery.

12.3.3 PHASE CHANGE OF SLAGS

Conventionally, solidified slags containing less than 5% crystalline phase are applicable as raw materials for the cement manufacturing industry. The primary crystalline phase in these slag melts was examined by XRD tests and the results are shown in Figure 6. It can be seen that the primary phase was $CaAl_2Si_2O_8$, and Ca_2SiO_4 was precipitated upon further decreasing the temperature, which agreed with the Factsage phase diagrams. The latent heat of phase change ($CaAl_2Si_2O_8$) was around 421.81 KJ/mol referred to Factsage software data [30]. The small amount of crystals (5%) formed in the melts could result in the increase of waste heat recovery due to the heat release from phase transformation, but it was difficult to control the crystallization degree in practical operation processes. It is therefore proposed that the liquid slags should be rapidly cooled to avoid crystallization according to the TTT curves.

12.3.4 INDUSTRIAL PROTOTYPE PLANT

Based on the aforementioned analysis, an industrial prototype plant was proposed for the purpose of recovering the high quality waste heat from high temperature slags, and the process flow is shown in Figure 7. The whole process flow could be divided into several parts. Firstly, SHTT experiments were carried out for a slag sample and the TTT curves were obtained. Secondly, the waste heat of the high temperature slags produced in steel industry was extracted with the dry granulation based on a multi-stage control method and the integrated heat was used by coal gasification, hydrogen production or heat storage in PCM. Then the solidified glassy slags could be utilized as cementitious material and finally both the waste heat and slag resources were recovered.

The possible recovered energy of this prototype plant based on multi-stage control method was calculated and listed in Table 2, in which average heat capacity of slags was referred to Factsage software data [30], the crystal content was assumed as 1% and the slags were assumed to be cooled to ambient temperature (25 °C). As can be seen, the calculation was divided into three stages based on the multi-stage control method. Accordingly the extraction of the high quality energy of slags could be divided into three stages, during which several operational parameters were controlled including heat exchange time and cooling rates; and finally different quantities of waste heat was recovered. For example, the possible extracted waste heat for sample A1 was 160.94 MJ, 198.91 MJ and 1235.86 MJ per ton of slags in the temperature range of 1500–1360 °C (Stage 1), 1360–1100 °C (Stage 2) and 1100–25 °C (Stage 3), respectively, and the total extracted waste heat could be up to 1695.71 MJ per ton of slags without phase change. Accordingly, the cooling rate and heat exchange time during stage 1 and 3 could be adjusted and that during stage 2 should be controlled larger than 6 °C/s and less than 43.3 s. With the heat release of 151.73 MJ per ton of slags from 1% crystal precipitation, the potential of total extracted waste heat could be more than 1847 MJ per ton of slags. In a similar way, the potential extracted waste heat during different stages for samples A2 and A3 could be calculated and the total extracted waste heat for samples A2 and A3 without phase transformation was 1793.82 MJ and 1828.82 MJ per ton of slags, respectively, as listed in Table 2.

12.4 CONCLUSIONS

The present study provided an analysis of high temperature slags for the purpose of recovering both the high quality waste heat and the slag resources. Three samples were designed containing different levels of Al_2O_3 at a fixed CaO/SiO_2 ratio of 1.05. A Single Hot Thermocouple Technique (SHTT) was used to investigate the slag properties during isothermal experiments and Time Temperature Transformation (TTT) curves were therefore obtained. The main conclusions were summarized as below:

1. A multi-stage control method of waste heat recovery from high temperature slags was proposed based on TTT curves, in which the whole temperature range was divided into three regions, i.e., Liquid region, Crystallization region and Solid region. Some significant technological parameters, including heat exchange time and cooling rate should be effectively controlled during these stages;
2. The polynomial expressions fitted the TTT curves well and could be used to deduce the incubation time at any temperature;
3. An industrial prototype plant was put forward aiming to recover the high quality waste heat of high temperature slags, and accordingly the potential of the extracted waste heat was calculated.

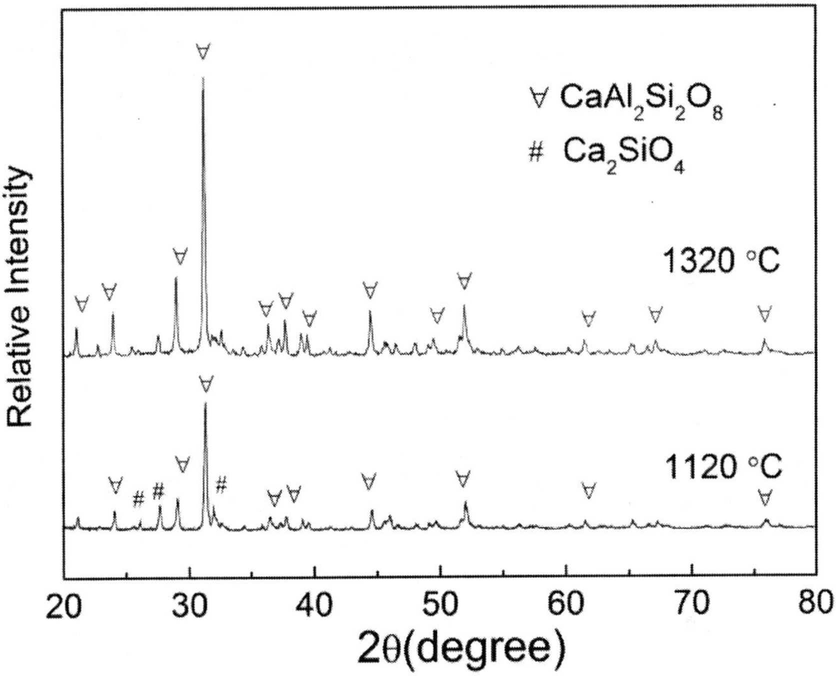

FIGURE 6: XRD patterns of quenched sample A2 from 1320 °C and 1120 °C.

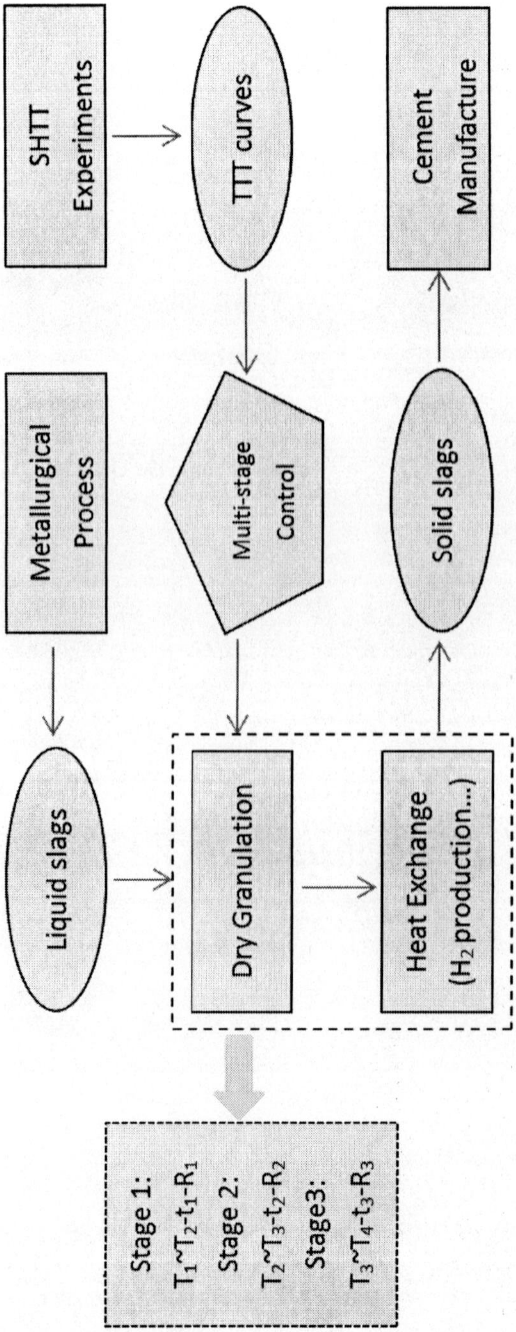

FIGURE 7: Process flow for the proposed prototype plant.

TABLE 2: Energy accounting of waste heat recovery based on multi-stage control.

Sample	Stage	Temperature range (°C)	Heat exchange time (s)	Cooling rate (°C/s)	Heat capacity (J/mol/K)	Possible recovered waste heat (MJ/t)	Phase change heat (MJ/t)	Total extracted heat (MJ/t)
A1	1	1500–1360	–	–	69.99	160.94	Crystals: 1%	Non–Crystallization: 1695.71
	2	1360–1100	43.3	6		298.91	151.73	Crystallization: 1847.44
	3	1100–25	–	–		1235.86		
A2	1	1500–1380	–	–	78.32	145.94	Crystals: 1%	Non–Crystallization: 1793.82
	2	1380–1100	18.7	15		340.52	151.73	Crystallization: 1945.55
	3	1100–25	–	–		1307.36		
A3	1	1500–1400	–	–	81.51	123.99	Crystals: 1%	Non–Crystallization: 1828.82
	2	1400–1080	6.4	50		396.76	151.73	Crystallization: 1980.55
	3	1080–25	–	–		1308.07		

REFERENCES

1. Milford, R.L.; Pauliuk, S.; Allwood, J.M.; Muller, D.B. The roles of energy and material efficiency in meeting steel industry CO2 targets. Environ. Sci. Technol. 2013, 47, 3455–3462.
2. Allwood, J.M.; Cullen, J.M.; Milford, R.L. Options for achieving a 50% cut in industrial carbon emissions by 2050. Environ. Sci. Technol. 2010, 44, 1888–1894.
3. Bisio, G. Energy recovery from molten slag and exploitation of the recovered energy. Energy 1997, 22, 501–509.
4. Barati, M.; Esfahani, S.; Utigard, T.A. Energy recovery from high temperature slags. Energy 2011, 36, 5440–5449.
5. November 2012 Crude Steel Production. Available online: http://www.worldsteel.org/media-centre/press-releases/2012/11-2012-crude-steel.html (accessed on 7 January 2014).
6. Zhang, H.; Wang, H.; Zhu, X.; Qiu, Y.J.; Li, K.; Chen, R.; Liao, Q. A review of waste heat recovery technologies towards molten slag in steel industry. Appl. Energy 2013, 112, 956–966.
7. Cai, J.J.; Wang, J.J.; Chen, C.X.; Lu, Z.W. Recovery of residual heat integrated steelworks. Iron Steel 2007, 42, 1–6.
8. China Iron and Steel Association. Available online: http://www.chinaisa.org.cn (accessed on 7 January 2014).
9. Pandelaers, L.; D'alfonso, A.; Jones, P.T.; Blanpain, B. A quantitative model for slag yard cooling. ISIJ Int. 2013, 53, 1106–1111.
10. Li, Q.H.; Meng, A.H.; Zhang, Y.G. Recovery Status and Prospect of Low-Grade Waste Energy in China. Proceedings of the International Conference on Sustainable Power Generation and Supply, Nanjing, China, 6–7 April 2009; pp. 1–6.
11. Van Oss, H.G.; Slag-iron and steel. US Geological Survey Minerals Yearbook; U.S. Department of the Interior and U.S. Geological Survey: Reston, VA, USA, 2010; pp. 69.1–69.10. Available online: http://minerals.usgs.gov/minerals/pubs/commodity/iron_&_steel_slag/index.html#myb (accessed on 7 January 2014).
12. Nakada, T.; Nakayama, H.; Fujii, K.; Iwahashi, T. Heat recovery in dry granulation of molten blast furnace slag. Energy Dev. Jpn. 1983, 55, 287–309.
13. Guo, H.; Zhou, S.H. Discussion about Heat Recovery Technology of Blast Furnace Slag. Proceedings of the Ironmaking Technology Conference and Ironmaking Academic Annual Meeting, Beijing, China, 26–28 May 2010; pp. 1141–1175.
14. Mizuochi, T.; Akiyama, T.; Shimada, T.; Kasai, E.; Yagi, J.I. Feasibility of rotary cup atomizer for slag granulation. ISIJ Int. 2001, 41, 1423–1428.
15. Pickering, S.J.; Hay, N.; Roylance, T.F.; Thomas, G.H. New process for dry granulation and heat recovery from molten blast furnace slag. Ironmak. Steelmak. 1985, 12, 14–21.
16. Akiyama, T.; Oikawa, K.; Shimada, T.; Kasai, E.; Yagi, J.I. Thermodynamic analysis of thermochemical recovery of high temperature wastes. ISIJ Int. 2000, 40, 286–291.
17. Shimada, T.; Kochura, V.; Akiyama, T.; Kasai, E.; Yagi, J.I. Effects of slag compositions on the rate of methane-steam reaction. ISIJ Int. 2001, 41, 111–115.

18. Kasai, E.; Kitajima, T.; Akiyama, T.; Yagi, J.I.; Saito, F. Rate of methane-steam reforming reaction on the surface of molten BF Slag: For heat recovery from molten slag by using a chemical reaction. ISIJ Int. 1997, 37, 1031–1036.
19. Yoshinaga, M.; Fujii, K.; Shigematsu, T.; Nakata, T. Method of dry granulation and solidification of molten blast furnace slag. Tetsu-To-Hagane 1981, 67, 917–924.
20. Sieverding, F. Heat recovery by dry granulation of blast furnace slags. Steel Times 1980, 208, 469–472.
21. Yoshida, H.; Nara, Y.; Nakatani, G.; Anazi, T.; Sato, H. The Technology of Slag Heat Recovery. Proceedings of the NKK SEAISI (Nippon Kokan KK-Southeast Asia Iron and Steel Institute) Conference of Energy Utilization in the Iron and Steel Industry, Singapore, 10–13 September 1984; pp. 1–20.
22. Xie, D.; Jahanshahi, S. Waste Heat Recovery from Molten Slags. Proceedings of the 4th International Congress on Science and Technology of Steelmaking, Gifu, Japan, 6–8 October 2008; pp. 674–677.
23. Purwanto, H.; Akiyama, T. Hydrogen production from biogas using hot slag. Int. J. Hydrog. Energy 2006, 31, 491–495.
24. Li, P.; Yu, Q.B.; Qin, Q.; Lei, W. Kinetics of CO2/Coal gasification in molten blast furnace slag. Ind. Eng. Chem. Res. 2012, 51, 15872–15883.
25. Li, P.; Yu, Q.B.; Xie, H.Q.; Qin, Q.; Wang, K. CO2 gasification rate analysis of Datong coal using slag granules as heat carrier for heat recovery from blast furnace slag by using a chemical reaction. Energy Fuel 2013, 27, 4810–4817.
26. Nomura, T.; Okinaka, N.; Akiyama, T. Technology of latent heat storage for high temperature application. ISIJ Int. 2010, 50, 1229–1239.
27. Yajima, K.; Matsuura, H.; Tsukihashi, F. Effect of simultaneous addition of Al2O3 and MgO on the liquidus of the CaO–SiO2–FeOx system with various oxygen partial pressures at 1573 K. ISIJ Int. 2010, 50, 191–194.
28. Kashiwaya, Y.; Cicutti, C.E.; Cramb, A.W.; Ishii, K. Development of double and dingle hot thermocouple technique for in site observation and measurement of mold slag crystallization. ISIJ Int. 1998, 38, 357–365.
29. Kashiwaya, Y.; Cicutti, C.E.; Cramb, A.W.; Ishii, K. An investigation of the crystallization of a continuous casting mold slag using the single hot thermocouple technique. ISIJ Int. 1998, 38, 348–356.
30. Bale, C.W.; Chartrand, P.; Degterov, S.A.; Eriksson, G.; Hack, K.; Ben Mahfoud, R. FactSage thermochemical software and databases. Calphad 2002, 26, 189–228.
31. Liu, H.; Wen, G.H.; Tang, P. Crystallization behaviors of mold fluxes containing Li2O using single hot thermocouple technique. ISIJ Int. 2009, 49, 843–850.
32. Kashiwaya, Y.; Nakauchi, T.; Pham, K.S.; Akiyama, S.; Ishii, K. Crystallization behaviors concerned with TTT and CCT diagrams of blast furnace slag using hot thermocouple technique. ISIJ Int. 2007, 47, 44–52.
33. Zhou, L.J.; Wang, W.L.; Liu, R.; Brian, G.; Thomas, B.G. Computational modeling of temperature, flow, and crystallization of mold slag during double hot thermocouple technique experiments. Metall. Mater. Trans. B 2013, 44, 1264–1279.

Author Notes

CHAPTER 1

Acknowledgments

Authors would like to express their gratefulness to all those who have participated in the research activities over the years. A special thanks to those industries that have strongly believed in the recycle approach and have stimulated the authors with new problems.

CHAPTER 2

Acknowledgments

R. D. Arancon thanks the Department of Chemistry of the Ateneo de Manila University in the Philippines for the wonderful opportunity to learn. Also, heartfelt thanks are due to Jhon Ralph Enterina (University of Alberta, Canada) and Jurgen Sanes (Simon Fraser University, Canada) for help with some articles. Carol Sze Ki LIN acknowledges the Biomass funding from the Ability R&D Energy Research Centre (AERC) at the School of Energy and Environment in the City University of Hong Kong. The authors are also grateful to the donation from the Coffee Concept (Hong Kong) Ltd. for the "Care for Our Planet" campaign, as well as a grant from the City University of Hong Kong (Project No. 7200248). C. S. K. Lin acknowledges the Industrial Technology Funding from the Innovation and Technology Commission (ITS/323/11) in Hong Kong. R. Luque gratefully acknowledges the Spanish MICINN for financial support via the concession of a RyC contract (ref: RYC–2009–04199) and funding under project CTQ2011–28954-C02-02. Consejeria de Ciencia e Innovacion, Junta de Andalucia is also gratefully acknowledged for funding project P10-FQM-6711. R. Luque is also indebted to Guohua Chen, the Department of Chemical and Biomolecular Engineering (CBME) and HKUST for the provision of a visiting professorship as Distinguished Engineering Fellow.

Conflict of Interest

None declared.

CHAPTER 4

Conflict of Interest

The authors declare that there is no conflict of interests regarding the publication of this paper.

Acknowledgments

The authors wish to express their gratitude to the "Ministerio de Ciencia e Innovación" (Project CTM2011-25762/TECNO) and "Junta de Andalucía" for providing financial support. Dr. Rincón wishes to thank the Ramón y Cajal Program (RYC-2011-08783 contract) from the Spanish Ministry of Economy and Competitiveness for providing financial support.

CHAPTER 6

Acknowledgments

The authors wish to thank the NM AIST and COSTECH for sponsoring of this research, Arusha city council for allowing us to use their facility during waste characterization, the laboratory of Energy of the University of Dar es Salaam for allowing utilization of their laboratory for waste analysis.

CHAPTER 7

Conflict of Interest

The authors declare no conflict of interest.

CHAPTER 8

Acknowledgments

The research was carried out in the frame of the Eco-Innovation Project Prowaste, financed by European agency for competitiveness and innovation (EACI). All the partners of the project are acknowledged for their useful discussion. Particular Acknowledgments are due to Ubaldo Spina and Andrea Di Tondo for the design of the bench.

Author Contributions

Alessandra Passaro was involved in project coordination and management; Alessandro Marseglia contributed to experimental activity and prototype design; Mariaenrcia Frigione contributed to experimental activity and data analysis; Alfonso Maffezzoli started this activity proposing the concept of RPL beam reinforcement with pultruded rods; Antonio Greco performed theoretical analysis of experimental data.

Conflict of Interest

The authors declare no conflict of interest.

CHAPTER 10

Acknowledgments

The authors wish to thank the AUTO21 Network Centre of Excellence, Mitacs, and the Natural Sciences and Engineering Research Council of Canada (NSERC) for their funding support in the development of this research.

Author Contributions

Lindsay Miller, as primary author, spearheaded and coordinated the preparation of the paper with the other authors. Lindsay Miller, Katie Soulliere, Susan Sawyer-Beaulieu and Simon Tseng shared in the background literature research, contributed to the scientific discussion and write-up of the first draft of the manuscript, as well as its subsequent revision based on reviewer feedback. Edwin Tam contributed to the review and edit of the first draft manuscript prior to its submission for first round review. All authors read and approved the final manuscript.

Conflicts of Interest

The authors declare no conflict of interest.

CHAPTER 11

Acknowledgments

Special thanks to Tom Raymond, PE, Landfill/Ashfill Manager of ecomaine and Kevin Roche, General Manager of ecomaine.

CHAPTER 12

The authors gratefully acknowledge financial support by the Common Development Fund of Beijing and the National Natural Science Foundation of China (51074009, and 51172001). Supports by the National High Technology Research and Development Program of China (863 Program, 2012AA06A114) and Key Projects in the National Science & Technology Pillar Program (2011BAB03B02) are also acknowledged.

Conflict of Interest

The authors declare no conflicts of interest.

Index

I

K

L

M

N

O

P

Q

R

X

Y